Creole Genesis:
The Bringier Family and Antebellum Plantation Life in Louisiana

Creole Genesis:
The Bringier Family and
Antebellum Plantation Life in Louisiana

Craig A. Bauer

University of Louisiana at Lafayette Press
2011

Cover Images: front top, Hermitage Plantation (photo by the author); front bottom, White Hall Plantation as depicted in a painting by Christophe Colomb (courtesy of The Historic New Orleans Collection, accession no. 1971.40); rear, Michel Doradou Bringier (courtesy of The Historic New Orleans Collection, accession no. 1970.11.35).

ISBN 13 (paper): 978-1-935754-07-7

© 2011 by University of Louisiana at Lafayette Press
All rights reserved

http://ulpress.org
University of Louisiana at Lafayette Press
P.O. Box 40831
Lafayette, LA 70504-0831

Printed on acid-free paper.

Library of Congress Cataloging-in-Publication Data

Bauer, Craig A.
Creole genesis : the Bringier family and antebellum plantation life in Louisiana / Craig A. Bauer.
p. cm.
Includes bibliographical references and index.
ISBN 978-1-935754-07-7 (pbk. : alk. paper)
1. Bringier family. 2. Creoles--Louisiana--History--19th century. 3. Creoles--Louisiana--Biography. 4. Plantation life--Louisiana--History--19th century. 5. Plantation owners--Louisiana--Biography. 6. Elite (Social sciences)--Louisiana--History--19th century. 7. Sugar growing--Louisiana--History--19th century. 8. Louisiana--History--1803-1865. I. Title.
CT274.B767B38 2011
929'.20973--dc23
2011031220

For my wife and daughter, Betsy and Charlotte
and
For my parents, Rudy and Joyce

Contents

Preface ix

Chapter 1
    A Louisiana Venture 1

Chapter 2
    A Creole Generation 13

Chapter 3
    Doradou and Aglae 25

Chapter 4
    The Hermitage 41

Chapter 5
    Life in the City 49

Chapter 6
    Land and Livelihood 59

Chapter 7
    Masters and Slaves 73

Chapter 8
    Conflict and Loss 95

Chapter 9
    The End of a Way of Life 119

Epilogue 143

Notes 149

Appendices 173

Bibliography 205

Index 211

Preface

Images of what life was like in the Old South have long fascinated readers and moviegoers in America. Over the years historians, fiction writers, genealogists, tour guides, docents, and movie and television directors and producers have all added paint and design to the canvas that today makes up the public's perception of what life was like on the great plantation estates of times past. Each of the painters has brought a different perspective and objective to the perceptual palate of the Old South. Historians, with their emphasis on factual accuracy, are at times at odds with the fiction writer and movie producer who are more interested in the dramatic and its ability to draw the interest of the public than in historical perspective and accuracy. Whether from the "moonlight-and-magnolias" romanticized vision of plantation life as has been presented in the fictionalized media or from a study of the historical tomes written about the antebellum period, there remains a strong interest among the general public in the details of this American society and culture of long ago.

Much of the interest in life in the Old South centers around the grand homes that a relatively few southerners enjoyed. Although many of the stately homes of the antebellum planter elite have been lost to the ravages of time, those that remain serve as an important link with the good and the bad of the Old South's history. In addition to being standing symbols of the much romanticized antebellum South, these old houses not only represent the wealth, aesthetic taste, culture, and lifestyles of their former owners but also the sweat, toil, skills, and talents of a workforce impelled by the institution of slavery to build and maintain the plantation manor house and to provide the labor that produced the estate's commercial crops. Their toil and sweat provided the planter and his family with the wealth and resources to live the lifestyle enjoyed by the elites at the top of the financial and social ladders in the antebellum South.

Among the most endearing and oldest surviving antebellum plantation homes in Louisiana is the Hermitage. Located on the east bank of the Mississippi River near the town of Darrow in Ascension Parish, the house is among Louisiana's most prominent historic nineteenth-century structures. Though smaller in size and stature than many of its surviving counterparts, the graceful structure has been called "a nearly perfect example of the vernacular Louisiana form."[1] Historically relevant as perhaps the oldest surviving example of Classical Revival architecture in the state and as the preeminent homestead of the Bringier family—one of the region's most prominent and influential clans of the late colonial and antebellum periods in Louisiana history—the Hermitage house today is a popular attraction for tourists and a premier example of historic restoration.

Until the Civil War and the devastating consequences it wrought to their fortunes and lifestyles, few, if any other, families in Louisiana possessed as a whole more wealth and influence in the state than the Bringier family. The Hermitage plantation functioned throughout most of the nineteenth century as the rural centerpiece and hub of the influential family's holdings and activities. However, it was by no means the only large estate and manor home in the region owned by the Bringiers. As noted by Mary Ann Sternberg in her charming survey *Along the River Road: Past and Present on Louisiana's Historic Byway*, of the many historic places to be found along the Great River Road corridor between New Orleans and Baton Rouge, just in the parishes of St. James and Ascension alone the Bringier family holdings included many of the areas' most prominent estates. In St. James Parish there were White Hall (Maison Blanche), Bagatelle, Colomb, and Union. In Ascension Parish the family's properties included Bocage, Hermitage, Houmas, Brulé, Tezcuco, Bowden, and Ashland. Though extensive, the list does not include the holdings family members maintained elsewhere in the state, such as Fashion plantation in St. Charles Parish and the family's large city-house Melpomene in New Orleans. Among the Bringiers' properties and manor homes, the Hermitage and Melpomene held positions of special interest and fondness among the members of the family. For decades the Hermitage, the country home of Doradou and Aglae Bringier, remained the centerpiece of both business and family activities of the Bringiers. Similarly, the family's large city home, Melpomene, served as the gathering place for Bringiers from all branches of the family tree while they spent time in New Orleans. As focal points of family interest, the Hermitage and Melpomene receive special attention in this work since the activities and happenings at these two places personify the lives and events of all of the Bringier family members, regardless of their homestead.

Today many of the houses named are gone, including White Hall, Union, Melpomene, and most recently Tezcuco.[2] Like so many of their contemporary structures, the ravishes of weather, fire, neglect, and the wrecking balls have forever taken from us some of the once stately homes of the Bringiers. Fortunately, several of the remaining family properties continue to survive as residential properties. Unique among these is Bagatelle Plantation, which was saved from destruction to make way for a cement plant in 1977 by being hauled over the levee and floated by barge upriver to Plaquemine Point in Iberville Parish. Of the surviving Bringier estates, at this time only the Hermitage remains open for public visitations.[3]

Visitors to the Hermitage experience the living conditions enjoyed by a relatively small segment of the population of the Old South. Those looking for more information about life among the white elite during the plantation era in the American South have a large library of published works to consult. However, until now, individuals desiring to learn more about the lives of the Bringiers were required to visit archive collections across Louisiana and the South or consult a multitude of publications to find published tidbits of information about the members of the family. Because of the prominence of Bringier family members in so many areas in the colonial and antebellum history of Louisiana, including business, agriculture, society, politics, sports, and military events, authors writing about these topics have often cited the Bringiers as examples of the elite class of individuals who dominated so much of the social, economic, and political cultures of the time. Two individuals, Richard Taylor and Duncan Farrar Kenner, who married into the family and became integral members of the influential clan have even had biographies published on their lives.[4] Yet, in spite of their prominence, much of the story of the powerful and talented Bringier family remains largely untold.

The story of the Bringiers in many ways is the history of the culture of the elites and power in the Old South. It is the story of the lifestyle of the planter class who dominated the social, political, economic, and military history of the antebellum period in the American South. The Bringier story is also the history of land use, culture, architecture, science, sport, and war in Louisiana's sugar country, as well as the chronicle of a dynamic family with its intertwined genealogy and rich collection of stories, legends, and lore that typified so much of the life in the South of long ago. In relation to the age of man, the ascendancy of the Bringiers, as that of the plantation South itself, was relatively brief. Beginning with the arrival of Marius Pons Bringier during the Spanish colonial period and ending for all practical purposes within a few decades after the end of the Civil War when their treasured properties passed out of the hands of the family, their dominance in the region lasted little more than three generations.

This work tells the story of the Bringiers during the time they dominated much of the life and culture in the sugar country of southeast Louisiana. As such, it is also an account of much of the general history of the area during a time when it enjoyed great wealth and influence. It is largely because of the events and culture of the bygone decades when prominent families, including the Bringiers, basked in the glow of the South's wealth and influence that there remains today an interest among the general public in the ever-decreasing number of remaining grand plantation houses, which have come to symbolize, correctly or not, the public perception of life in old Louisiana. This study provides a unique multi-generational look at life among the powerful and influential planter elite in Louisiana from colonial times to the difficult and fractious times following the South's defeat in the Civil War.

Even though primary sources were consulted for data, or when possible to provide additional verification for information found in family notes and reminisces, this work is intended to recount the Bringiers' story as told and recorded by family members who captured first-hand memories of individuals who actually experienced the events discussed. Though nearly all of the family members examined in this work died before the twentieth century, their stories were not lost due to the effort of one individual, Trist Wood. A great-grandson of Doradou and Aglae Bringier, Wood spent decades before his death in 1952 collecting and writing down first-hand accounts of the Bringier family given by his aunts, uncles, and other family members. He also managed to collect hundreds of original copies of wills, deeds, inventories, letters, newspaper articles, and other papers relating to the many branches of his family tree, which include, along with the Bringiers, the DuBourg, Wood, Crooke, Dabney, Jennings, Taylor, and Trist families of Louisiana, Virginia, and Rhode Island.

Wood was more than just his family's genealogist. He was a fascinating character who possessed an interesting history of his own. The son of Robert Crooke Wood, a former city councilman in New Orleans and a Confederate veteran, and Wilhelmine Trist, he was a person of many talents and interest. As a young man Wood spent time in Europe, where he worked as an artist and editor of the monthly magazine *The Quartier Latin*. Upon his return to the United States, he worked for some time as a freelance artist and later as the cartoonist for the New Orleans *Item*. In his cartoons, Wood frequently sketched Louisiana's politicians with a mocking and bitterly sarcastic style. Although sometimes a target of Wood's drawings and sarcasm, Huey Long admired the work of the artist and hired him away from the *Item* by doubling his salary. Huey put Wood to work on his political publications including the *Louisiana Progress*, where his work was published weekly in cartoons depicting Huey as the hero of the people and his enemies in cruel caricature.[5]

Although Wood's collection is voluminous and encompassing, only a portion of it applies to the Bringiers. As is sometimes the case with family collections, Wood's papers dealing with the Bringiers are often uneven in their coverage of individuals and events. Issues of interest to Wood receive greater attention and detail while other issues and events are ignored. The result is at times an uneven look at the people and events of the period covered in this work. However, in spite of the spotty unevenness of his information, Wood's thousands of pages of material provide a revealing and fascinating peek into life in Louisiana during a pivotal period in its history. Not unexpectedly, as a son of the postbellum era in Louisiana and the South, much of Wood's recollections and materials reflect the popular themes of the period that promoted the romantic view of the antebellum period and the heroic struggle of white Southerners for the Lost Cause. Hence, details of the trials and tribulations of members of the family during the war are plentiful, while other important events and times (including whole years) go unmentioned. Reflective of the time and attitudes when he was collecting his notes, the most important topic largely ignored by Wood was insight into the many African American individuals and families who lived and worked on the Bringiers' properties. With the exception of brief mentions of servants or references that relate to business transactions—property inventories or slave purchases and sales—or to some member of the family or a family event, scant notice is found in Wood's papers of the hundreds of African Americans who resided and worked on the Bringier plantations. To address the omission, materials from different sources are used to shed light on the contributions made by the people whose sweat and toil contributed so much to the history of the Bringiers and the region.

Today, the bulk of the original copies of the Trist Wood Papers are owned by the Hermitage Foundation and are housed with the Foundation and at The Historic New Orleans Collection. A smaller portion of the papers is part of the Southern Historical Collection housed in the Manuscript Department at the Library of the University of North

Carolina at Chapel Hill. In the researching and writing of this work and an earlier study of Duncan Farrar Kenner, I was fortunate to have been able to work with the Trist Wood Collection while it was housed at the Special Collections Division of the Howard Tilton Library of Tulane University in New Orleans. Additionally, extensive use was made of a typescript of Trist Wood's papers meticulously prepared by Rose Warren and provided to me by Dr. Robert Judice. I extend my sincere thanks to Doctor and Mrs. Judice and the Hermitage Foundation who provided me access to their valuable collections of Bringier family materials, which were essential to the completion of this work. I am also indebted to Mr. Duke Rivet of the Hermitage Foundation's Board for the sharing of materials and for his comments on the draft of this manuscript. Special gratitude is extended to the late Dr. Jessie Poesch, also a member of the Hermitage Foundation Board, for her kind encouragement, assistance, support, and advice during the years it took to complete this project. I am also grateful for the research assistance provided to me by Sister Helen Fontenot, M.S.C. and Diana Schaubhut of the Blaine S. Kern Library at Our Lady of Holy Cross College and to the college itself for the awarding of the Nancy O'Neill Endowed Professorship III and the Freeport-McMoran Endowed Professorship in Business and Education that were instrumental in providing financial support for completion of this project. And finally, heartfelt appreciation and gratitude are extended to my wife, Betsy, and daughter, Charlotte, for their patience, support, and assistance in the researching and writing of this work. I hope that this story will add additional pieces to the grand mosaic of what life was like on the plantations that once flourished during a time and age long passed in the history of Louisiana.

Craig A. Bauer

Chapter 1

A Louisiana Venture

The year 1783 marked for a second time in less than three years that thirty-one-year-old Marius Pons Bringier and his young wife, Francoise Durand, abandoned a home and set sail on a vessel for a new land and beginning. Carrying not only their belongings with them but also a desire and hope for a new and happier life, the young couple bid farewell to the picturesque Caribbean island of Martinique. The Bringiers had moved to the French possession in 1781 with the hope of putting behind them unhappy events and memories from their time together in their native France.[6]

Born in France on October 27, 1752, Marius Pons lived in his native country until he was almost thirty. As an adult, he served as proprietor of his family's estate Lacadiere (La Cadiere), located near the town of Aubagne in southern France. There he lived with his wife, Francoise, and looked forward to having a family and a prosperous future. Unfortunately, their hopes and expectations failed to materialize. In spite of their wealth and stature, their life in France was not a happy one. The young couple wanted a family. Tragically, when their first child—a daughter named Marie Catherine—was born, she lived only a few weeks. The pain caused by the devastating loss of their child along with the mounting financial uncertainties caused by the declining social and economic situation in France associated with its calamitous colonial and military ventures, led Marius Pons and Francoise to look for a new beginning in a new land. Giving up their life at Lacadiere, the young couple decided to strike out for the New World. In spite of the increased dangers of crossing the Atlantic while France and Great Britain were at war due to France's involvement in the American Revolution, the young French couple willingly accepted the dangerous challenges of an ocean voyage. Sailing from the port of Marseilles in 1781, the couple chose the picturesque Caribbean island of Martinique as the place for their new beginning and their new home.[7]

Unfortunately, life on the island did not turn out as the young couple anticipated. Personal and business problems followed Marius Pons and Francoise to their new home. The joys of the births of two more children were again tragically followed by the deaths of both infants within days of the births. The effects of these tragic losses upon the couple were compounded when the business partnership between Marius Pons and his brother Vincent failed.[8] Faced with both the personal tragedy of the loss of their babies and their hope for a successful business venture shattered, Marius Pons and Francoise once again made the difficult decision to leave their problems behind to search elsewhere for a new start and a new hope for happiness and financial prosperity. Selling his share of his Martinique estate, Marius Pons and his wife—for the second time in less than three years—set out to establish a new life and home in a strange new land.[9]

The Bringiers looked now to the French community in Louisiana for their new home and hopes for a happier and more prosperous future. It is unclear what the attraction of the Louisiana colony was that drew Marius Pons and his wife to its shores in 1783. Although culturally and socially it remained predominantly French, the colony—at the time of the Bringiers' arrival—belonged to Spain and offered an unimpressive history of economic successes and opportunities to newcomers. Though the Spanish colony dwarfed their previous island home in land size, population wise the tiny island of Martinique enjoyed an advantage of nearly three to one. Perhaps it was Marius Pons' business acumen which led him to realize that with the end of the American Revolution, Louisiana and its mighty Mississippi River offered new and grand opportunities for economic prosperity, or perhaps it was the fact that Spanish officials were offering generous land grants to new immigrants which drew the Bringiers to Louisiana. Regardless, upon their arrival in the Spanish colony, the Bringiers went to work building a new life. Since Spanish officials based the size of their land grants upon the means of settlers, colonial officials enthusiastically received the affluent Bringiers' request for land. Marius Pons petitioned for and received a generous grant of land situated on the east bank of the Mississippi River a few miles upriver from New Orleans. The Bringiers were able to work immediately on building their new homestead since they brought with them both their household furnishings and their slaves from Martinique. Marius Pons acquired property near New Orleans at modern-day Carrollton (then called the Chapitoulas or Tchoupitoulas District). Unfortunately, the run of bad luck that had plagued them for so long continued. The Tchoupitoulas site proved to be a poor choice for a plantation. Troubled by all too frequent crevasses, or breaks in the primitive levee system, the resultant flooding caused by the levee breaches, and a recurrent issue of ill health among his slaves, Marius Pons decided to relocate his home and family one more time.[10]

The most detailed record of the experience comes from an account of one of Marius Pons' slaves named Augustin[11] who, years after the event, dictated the story to a member of the family. The old slave recalled that:

> My old master was a Frenchman . . . from France. . . . He came to this country in his own ship. Yes, it belonged to himself. He has his wife, but no children. As soon as he arrived he bought land near the city together with a band of Negroes who had come from their country. There were men, women, and children. My mother was one of these women. When he went to work the land there, he found it too low—nothing but palmetto covered it. Now, when the river was high, it drowned the road. Many Negroes died. It was said because they drank river water. There were some too who grieved when they thought of their country. Well, it disgusted old master. He left that place and bought another plantation further up.[12]

Marius Pons' final move took him and his family up river to an area called the Paroisse St. Jacques de Cabahanoce aux Acadians. In this area, which eventually became St. James Parish, Marius Pons found the land and soil conditions superior to his holdings near New Orleans. Excited over the potential offered by the area, the French émigré purchased five-adjoining plots of land on the east bank of the Mississippi River over the four-year period from 1785 to 1789.[13] Joining the plots together, Marius Pons named his new estate La Maison Blanche. However, because Marius Pons and Francoise found the English equivalent to be quaint sounding and unique in the area, which was dominated by French inhabitants and place names, they preferred to call their home "White Hall."[14]

Being involved in both large-scale-agricultural production and the shipping business with his sailing vessel, Marius Pons soon experienced the financial success and status he had long sought. Taking advantage of the many business opportunities available in Louisiana, including increased trade with the newly independent American states, Marius Pons rapidly increased his personal wealth. Though both the Spanish officials and French residents of Louisiana looked upon the rising tide of American commerce and immigration with more than a little suspicion, it was obvious to Marius Pons and others that the situation brought opportunity as well as problems to the region. In spite of official efforts by the Spanish authorities in Spain and Mexico to limit the presence of the Americans in the colony, local residents and authorities often disregarded official Spanish policy and carried on business with the Americans. Marius Pons' agricultural and business endeavors proved so successful that he soon amassed a fortune that allowed him and his family to live a lifestyle few in colonial Louisiana could match.

Throughout much of the colonial period in Louisiana history, residents of the area built houses that were fairly simple in style and plan. At the time, French influence played a major role in the design of most early rural homes in the colony. Most houses were small, consisting usually of two or three rooms in a row. The houses were typically built with galleries, supported with wooden colonettes, across the front and frequently the back of the structure. Other common features of Louisiana's early homes included broad-spreading rooflines and, if a larger structure, multiple French doors. The typical construction materials included heavy timber frames, usually of cypress wood, and fill between the timbers of either a mixture of mud, Spanish moss, and animal hair called *bousillage* or of locally made bricks (*briquette entre poteaux*). When the mud fill was used, wood bars called *barreaux* were set between the posts to hold the plaster in place. To protect the support materials from weather damage, the walls of the structure were covered either with plaster that was later painted or with horizontal cypress boards (*madriers*). Hand-rived cypress shingles covered the roof and all windows and outer doorways were shuttered. Although some of the wealthier planters built sizeable plantation homes in the region as early as the 1720s, these structures were more or less larger versions of the simple rural home. In addition to size, the major distinction of the early plantation houses was that they were frequently raised approximately one story off the ground. The main floor was on the second level and usually protected by a gallery. The raised living quarters provided the family with a storage space below and, more importantly, provided the family home with protection from dampness and the more damaging flooding that frequented the area because of the lack of an effective levee system. Marius Pons wanted more.[15]

Foremost among his personal goals was the construction of a home, which would denote to all his standing, taste, and growing prosperity. He was among the earliest planters in the region to have the resources and dedication to abandon the more utilitarian structures of the time for a truly grand home. Toward his goal, Marius Pons constructed one of the region's most notable manor homes—White Hall. With its white pillars and walls, from which its name was derived, and its Italian-villa architectural design of Roman arches, rustication, balustrades, and urns, the house contrasted vividly with the older-style and simpler plantation homes of the region. An elaborate iron fence—which according to family tradition was imported from Europe—further enhanced the appearance of the imposing mansion. Regrettably, with the exception of a painting by Marius Pons' son-in-law Christophe Colomb, few records or renderings of the imposing mansion remain today. Christophe's picture shows a building similar in many ways in design and style of the Cabildo in New Orleans surrounded by a wooden fence. The structure included five large arched bays on each level and walls that were probably veneered with plaster and were designed to look like grayish-white marble.[16] The design, with the addition of an ornamental frieze, gave the house an appearance similar to that found in the early design of the Cabildo. The likeness in style between the two colonial structures has led some to speculate that Marius Pons' house was possibly the work of Gilberto Guillemard, the designer of the regal Spanish government building located next to St. Louis Cathedral on the square in New Orleans. Both floors of the two-story structure were divided into three large chambers with a central area on the first floor serving as a salon that held the main staircase to the upper level of the house. A second set of stairs to the upper floor was located within the rear gallery of the house. Black and white marble walkways, well-kept groves, orchards, luxurious gardens, and artificial ponds surrounded the grand house. Inside, the home was filled with expensive pieces of furniture and the walls were decorated with costly works of art throughout the mansion. Such magnificence soon earned the estate a reputation at the time of being among the fairest in the South.[17] Such were the thoughts of Pierre Clement de Laussat, the French Commissioner and former Prefect of Louisiana, who upon visiting Marius Pons' estate noted that:

> The house he [Marius Pons] built for himself is the most substantial, the best constructed, best appointed, and most distinguished of the country houses in the colony. The roof forms an Italian terrace with a balustrade all around.... The exterior buildings speak of work, industry, and affluence, here are seen corn and cotton mills, presses, etc. These mills are run by horses.[18]

Though impressed with the structure and operation of the plantation, the French commissioner was somewhat less complimentary of his host's decorating talents, observing that "the interior arrangement [at White Hall] lacks taste and comfort."[19] Nevertheless, in spite of the acerbic opinion of Laussat concerning the interior of the manor house, its imposing presence on the banks of the Mississippi during Louisiana's colonial period qualifies it as one of the most unique historic homes ever constructed in Louisiana.

Over time, Marius Pons continued to expand the size of his estate. By the time of his death in 1821, his White Hall plantation had grown to more than twenty-seven arpents[20] fronting on the river by forty arpents deep. As he expanded his holdings, Marius Pons also added important utility structures on the property including a thirty-six-foot square two-story brick house, two pigeon houses with rooms underneath, a milk house and brick oven, a stable, and a flour mill. A shed for drying bricks that were made on the plantation, a sheepfold, a forge, and a cotton gin and press also were located on the property. Animal stock on the estate included a flock of approximately 160 sheep, thirty milch cows, several oxen, mares, and saddle horses.[21]

Not all the memories of their former home on Martinique were negative for the Bringiers. While a resident of the little island, Marius Pons and Françoise developed a love of the many beautiful birds and flowers that inhabited the Caribbean island. Hoping to import some of the beauty they had enjoyed on the tropical island, the Bringiers took elaborate steps to recreate the plush beauty of their former island home in Louisiana. Though an avid hunter, Marius Pons took the unusual action for the time of taking complex measures to preserve the environment and natural beauty of his land. To accomplish this rare effort at environmental awareness, he constructed a system at White Hall of intricate hothouses and aviaries to hold collections of rare specimens of flowers and birds procured from around the world. He even placed large areas of his property off limits to the discharging of firearms to protect the native animals and birds of the area. Birds in his preserves could only be captured by using netting. Additionally, a superb kitchen garden and an orchard with a great variety and number of fruit trees, including lemon, orange, and guava were added to the trees on the estate. To be able to better highlight their beauty and aesthetic effect and to protect them from bad weather and other mischief, Marius Pons had the tender fruit trees planted in unique large boxes, which were equipped with wheels that permitted them to be moved on special rails around the plantation. Also on the estate were several large ponds, which he kept stocked with varieties of his favorite fish and turtles.²²

Having learned from his earlier failed effort at plantation building at Tchoupitoulas, Marius Pons was also careful to build good housing and a hospital for his slave force at White Hall. Family tradition maintained that Marius Pons was a fair and kindly master to his slaves. Whether the members of his slave force of fifty-two agreed with this characterization is not known. Relatively little is known of Marius Pons' slave force in St. James Parish, with the exception of the fact that he maintained a hospital on the estate for his slaves—a fairly common practice on larger plantations of the time—and the fact that among the workforce were men and women who were considered to be good servants and others who were skilled tradesmen. Furthermore, there is no evidence that any of the Bringier slaves participated in the large and bloody slave insurrection that occurred in 1811 in St. John the Baptist Parish just a few miles down river from the Bringier plantation.²³

As was the case on many of the other agricultural estates operating in southeast Louisiana during most of the colonial period, the primary crops grown on the Bringier plantation were indigo, cotton, and rice. Sugarcane was also grown in the area during the colonial period, but because the product was wet, partially granulated, and leaked from casks so badly, it was not a viable product to be produced on a large scale. Although the arrival of the cotton gin led Marius Pons to grow cotton for a while, once Etienne de Boré solved the sugar granulation problem in 1795, production of the ribbon crop across the region expanded quickly and eventually became the major money crop on the White Hall estate as well as on most of the other large agricultural units in southeast Louisiana.²⁴

Having been an individual of substance in France and Martinique, Marius Pons was fully aware of the many benefits that having friends in high places brought to the individual in both Europe and the colonies. Though his wealth and financial success alone were enough to win him special consideration in early Louisiana, Marius Pons was also determined to become an active and contributing citizen in his new land and a loyal subject of the Spanish crown. Shortly after he moved from the Tchoupitoulas district to St. James Parish, he joined the local unit of the colony's militia. Originally given the rank of first lieutenant in the Provincial Infantry Regiment of the German Coast, he enjoyed his militia service and continued to serve in the Spanish colonial militia for years, eventually rising to the rank of captain.²⁵

Though he enjoyed his service in the militia and had a great fondness for the exotic birds and plant life found on his estate, Marius Pons was at his happiest when he entertained guests at his impressive home. A fervent adherent of the Southern tradition of cordial hospitality, he was always happy to welcome both expected and unexpected guests to his family's White Hall home. His open-arms hospitality earned the Bringiers a widespread reputation for their lavish parties and elaborate meals. According to family tradition, travelers who stopped at the estate were offered a ready welcome with no questions asked of their identity by the hosts. After their stay at the opulent plantation home, which was provided at no cost to themselves, visitors were free to proceed on their way with their identities undisclosed.[26]

Visitors to the Bringier house ranged from ordinary people of the area to some of the most powerful and famous people to visit south Louisiana during the period. Among the more celebrated visitors who took advantage of the Bringier family's hospitality was the naturalist John James Audubon[27] and a future King of France. In 1797, the then exiled Duke d'Orleans, Louis Philippe, was treated to the culinary and social delights for which White Hall had become renowned. So memorable was the visit that decades later when he obtained the French throne in 1830, the French monarch, in gratitude for the gracious hospitality shown to him during his visit to their home, sent the Bringiers a set of handsome china plates. The dinnerware remained among the family's most cherished possessions and was solemnly handed down from one Bringier generation to another.[28]

In spite of the memorable nature of the French nobleman's visit, the White Hall houseguests most honored and celebrated by the Bringiers were General and Mrs. Andrew Jackson. Family tradition maintains that the Bringiers hosted the hero of the Battle of New Orleans and his wife, Rachel, at White Hall sometime after the battle. It is unclear as to how the sophisticated French immigrant and the rough frontier Indian-fighter from Tennessee became acquainted. It is possible the contact between the two men came from Marius Pons' son Doradou, who served as a volunteer aid to the American general during the campaign against the British. However, it is more likely the relationship developed through the auspices of either the politically powerful Martin Gordon—a leading Jackson supporter and acquaintance of the Bringiers—or those of the Catholic prelate Abbé DuBourg—the Apostolic Administrator in New Orleans who was instrumental in honoring the general after the American victory over the British and was the uncle of Marius Pons' daughter-in-law Aglae. Considering the differences in cultural backgrounds of the Jacksons and Bringiers, it is not surprising that among the most notable remembrances of the historic visit passed down through Bringier family tradition concern the famous frontier couple's etiquette. According to family lore, the refined and gallant manners of the general and the lack of polish and sophistication of his wife intrigued the Bringiers. Though never vulgar or unladylike, Rachel Jackson was said to have made a lasting impression upon the Bringiers by her practices of calling the ladies of White Hall "honey" and her frontier custom of smoking a corncob pipe. Because the story of the future president's visit is based on family legend, the actual date of the visit is unclear. Jackson's publicized postwar visits to New Orleans to celebrate anniversaries of the momentous battle at Chalmette in 1828 and in 1840 do not appear to be plausible dates for a Jackson stay at White Hall. If 1828 was the date of the visit to the Bringier home, as some believe, Marius Pons could not have been present for he was dead for nearly eight years by time the general made his celebrated return to the Crescent City at the start of his successful campaign for the presidency. The 1840 visit of the ex-president to New Orleans is even a less plausible date for the visit to have occurred at White Hall, for Rachel had died shortly after Jackson's election to the presidency in 1828 and the once elegant plantation house was vacant and largely abandoned by that time. However, it was during his 1840 visit to the area to mark the fortieth anniversary of his victory over the British that Jackson did make a brief visit to Doradou Bringier's Hermitage home. Hence, the only plausible time for the Jacksons to have visited White Hall together was during the several weeks following the battle in 1815 when Rachel and the couple's adopted son, Andrew Jr., joined the general for an almost month-long stay in the region.[29]

Happily for the Bringier couple, the tragic losses of their newly born children in France and Martinique were not repeated in Louisiana. Six children were produced from the marriage after the couple settled in Louisiana.[30] Though all six of the Bringier children born in Louisiana survived to adulthood, their mother, Marie Francoise Durand, did not live long enough to enjoy much of the wealth and splendor the family acquired in Louisiana. In late spring 1803 as the United States was in the process of completing arrangements for the acquisition of Louisiana from France, she died of unreported causes leaving several young children to be reared by their father and his servants.[31]

As was common practice at the time for widowers, Marius Pons did not wait long before remarrying. Within a year of the death of his beloved first wife, he married Marie Anne Roudanez on April 5, 1804. Known to the Bringiers as "Nanene," she was a native of San Domingo who had narrowly escaped with her life during the slave uprising on that island. In Louisiana, she resided near the Bringier home with a servant named Nerestine Roudanez. In spite of Nerestine's slave status, the two women were very close. A lifelong relationship developed between the two women at the time of the slave uprising when Nerestine risked her safety and freedom to save Nanene's life by helping her escape San Domingo. According to family tradition, Nanene's marriage to Marius Pons was more of convenience than of love. Bringier lore holds that perhaps the main reason for the marriage was Marius Pons' belief that she would be an excellent mother for his younger children. The business character of the formal marriage contract signed by the couple the day before their wedding supports the family's interpretation of the relationship. In the agreement, Nanene forfeited her claim to community property with her husband in exchange for benefits after his death, which included a single $800 payment, interest on $8,000, and various pieces of furniture.[32]

Marius Pons and Nanene remained together for more than sixteen years until Marius Pons' death at age sixty-seven on April 21, 1820. Shortly after the death of her husband, Nanene moved from the White Hall manor house to two small houses located on the estate not far from the home she had shared with Marius Pons. Since she was the only member of her family to have survived the slave revolt on her home island of San Domingo, Nanene's remaining years were spent largely in seclusion having few visitors except for her lifelong attendant Nerestine, her other servants, and on occasion members of the Bringier family.[33]

With the exception that death came while he was at White Hall, little is known of the cause of Marius Pons' passing. Originally buried in the Catholic churchyard cemetery in St. James Parish, his body had to be disinterred and moved to another site because of the river's encroachment on the original burial site. Marius Pons' remains and several other family members who were buried in the St. James Cemetery were eventually removed to the grand family tomb at the Catholic Cemetery in Donaldsonville.[34]

The transfer of Marius Pons' remains to the family's Donaldsonville tomb occasioned some of the more bizarre stories in the Bringier family's colorful history. No doubt the reinterment process, which for an unknown reason required the moldered and partially decayed corpse of Marius Pons to be laid out for all to view for a time at the Hermitage home of his son Michel Doradou while final preparations were made at the new tomb, provided grist for the plantation's storytellers for decades after the event. Following his internment, slaves at the Hermitage plantation maintained that a visit to the Bringier tomb in the dead of night would reveal the large stone urn, which crowned the summit of the tomb, open, shooting forth flames. In the midst of the smoke and flames, they claimed that their old master Marius Pons could be seen seated at a table gambling with his son-in-law Christophe Colomb as they had often done during their lifetimes![35]

Perhaps an even stranger tale, because it no doubt was closer to reality than the slaves' lore of the flaming urn, was the story of the purloined teeth. In his great collection of notes and documentation of the Bringiers' history, Trist Wood reported a story passed directly onto him by his aunt and Marius Pons' granddaughter, Stella Bringier. According to the account, during the time that the partially decayed remains were on display at the Hermitage, an individual, identified by Wood simply as "a descendant of one of his daughters," admired the fine teeth of the corpse that remained intact in the skull. Suffering from a lack of his own incisors and being ever practical, the nearly toothless relative had a slave surreptitiously remove the objects of their admiration in the dead of night. The teeth were taken to New Orleans where a dentist fashioned them into dentures for the toothless relative![36]

After the death of Marius Pons, the once magnificent White Hall passed into the hands of others. As called for in the terms of his will, the estate was put up for sale. For a time, the property remained in the Bringier family since Marius Pons' son Doradou purchased it. In spite of deep sentimental feelings for his family's home, Doradou found managing both estates—Hermitage and White Hall—financially and operationally difficult. Doradou eventually sold White Hall to the wealthy South Carolinian Wade Hampton, who had recently brought a large slave force to Louisiana and was buying up properties along the Mississippi in a successful effort to establish a profitable sugar plantation.[37]

Surprisingly, Wade Hampton's acquisition of White Hall did not close the chapter of the Bringier family's interest and involvement with the property in St. James Parish. Obviously not all the Bringiers were comfortable with the passing of the family's homestead from their grasp. Within months of his death in 1847, Doradou Bringier's widow reacquired the estate in 1848 from the Hampton family and placed her son M. S. (Marius Ste. Colombe) Bringier in charge of the property.

Unfortunately, the family's dreams to restore Marius Pons' magnificent White Hall manor as a family centerpiece went unfulfilled. Approximately two years after Mrs. Bringier regained the property, the once grand manor house was irreparably damaged by fire. With only several small outbuildings, the sugarhouse, and the ruins of the mansion remaining, M. S. Bringier acquired the property from his mother in February 1854 and primarily used the area as a stock farm.[38]

Chapter 2

A Creole Generation

In spite of the destruction of his White Hall estate, Marius Pons Bringier's legacy of grandeur and accomplishment continued on long after his death and the ruin of his stately home. All of Marius Pons' six children, who lived to adulthood, followed in the footsteps of their father's success and achieved in their lifetimes levels of social and economic status which placed them among Louisiana's leading Creole[39] families and at the very pinnacle of the ruling class in early Louisiana. The oldest of Marius Pons' children was a son, Paul Louis. Known all of his life by the praenomen of Louis, he was born in America on his father's plantation at the Tchoupitoulas district on August 25, 1784. Shortly after his birth, the family moved to White Hall, their new home in St. James Parish. Throughout his life, Louis exhibited a flare for life and often demonstrated a non-traditional, wild, and sensational lifestyle that sometimes resulted in great success and good fortune and other times prodigious failure. An example of Louis' luck and timely fortune occurred when he was just a youth and his father played host at White Hall to the Spanish governor-general and his official entourage of more than one hundred individuals. As was the tradition at White Hall, the governor, Baron de Carondelet, was treated sumptuously while he and his cortege were at the St. James' plantation. Their host generously provided for even his cavalry escort. At the end of the visit, the governor directed that his assistants reimburse Marius Pons for his expenses, particularly those involved with providing for his military entourage. When Bringier politely, yet firmly refused to take any compensation for his hospitality, the governor, being equally determined that his host accept some gratuity for his effort, insisted that Marius Pons accept something as a sign of his appreciation. The matter was settled when the Spanish official directed his secretary to draw up a deed conveying a large tract of land in what is now northern Louisiana to his host's eldest son, Louis, who was present during the discussion. The generous reward of forty-thousand arpents of land—located at the junction of the "Rio Negro" (the Black River) and Bayou Tensas—was the first land grant given by the Spanish in that largely desolate and unpopulated region of the colony. However, with their efforts focused on their holdings along the lower Mississippi, neither Marius Pons nor the young Louis made any effort to settle the area at the time of the gift.[40]

As the family's first son, Louis was given management responsibilities as he grew older. Unfortunately, the fun-loving and adventurous Louis was not always up to the tasks assigned by his father. For example, when only age sixteen, his father assigned him the responsibility of traveling alone to New Orleans to handle some relatively minor financial matters. While there, the young man from the rural countryside found the illicit attractions of the Crescent City more than he could handle. Louis soon forgot about his duties and went on a spree that ended only when he had spent and lost all the money he had brought with him by gambling and enjoying other enticements the city had to offer a young man on his first solo business venture. When he realized what he had done, young Louis was ashamed and apprehensive about returning home and reporting to his father the results of his failed mission. Fortunately for Louis, an older friend, Christophe Colomb, who possessed a calming and diplomatic way, agreed to help the young man work through his problem with his father. The two returned to White Hall where Marius Pons took an immediate liking to the suave Colomb and all was forgiven. So harmonious was the meeting and the elder Bringier's liking for Louis' friend that he was permitted to remain at White Hall for an extended stay. During his long visit at the Bringier's plantation, the debonair Colomb spread his charm among the other family members to such an extent that he managed to win the hand of his host's eldest daughter Francoise (Fanny).[41]

Unfortunately, the lesson learned by Louis from his misadventure with the gambling establishments of New Orleans was only temporary. In 1807, several years after his first debacle and at a time when he had successfully established his own estate at the former Andry plantation on the east bank of the Mississippi River in Ascension Parish, Louis again became infected with the gambler's fever. As with his first ill-fated business venture in New Orleans, Louis was once again charged by his father with a mission of disposing of several barges loaded with cotton and indigo produced at his family's plantations. As he usually did when he visited the city, young Louis spent time with several friends who lived in New Orleans. As part of their night-out activities, they visited the city's gambling establishments where Louis' luck was extremely good and he won a large sum of money. His friends persuaded him to end his play before his luck changed and accompanied him back to his room and said goodnight. Excited by his good fortune, Louis could not sleep. He later told relatives that he was haunted by the thought that his luck was running so good that if he had continued to play he would have bankrupted the gambling house. Unbeknown to his friends, he got up, dressed, and returned to the casino. Unfortunately, this time his luck turned and he quickly lost all of his winnings. Desperate to see his good luck return, he continued to play. Running out of his own funds, Louis even wagered the proceeds of his father's cotton and indigo sales—and lost all. Humiliated, ashamed of his folly, and unwilling to return for a second time to report his misadventure at a New Orleans' casino to his father, Louis abandoned his life and family and simply disappeared from Louisiana without a word of explanation to his relatives and friends. For years family members had few clues as to what had become of him.[42]

Years later the family learned that after his departure from New Orleans, Louis traveled north and spent time in the Arkansas and Missouri territories. Though short in height, he was broadly shouldered and possessed the same extraordinary body strength that both his father and brother were known to have. His excellent physical condition stood him well in his adventures away from Louisiana. While in the frontier regions to the north of his native territory, he lived a life very different from the seigniorial one he had enjoyed on his family's plantations along the lower Mississippi River. Living and trading with several of the region's Indian tribes that inhabited the frontier, including the Osages, Louis spent years exploring the unchartered region and became an explorer of some note. In 1812, he was an eyewitness to the devastating New Madrid earthquakes and became the first individual of European descent to discover the Toltec Mounds in Lonoke County, Arkansas—a future National Historic Landmark. Although written several years after the event, his account of the devastating earthquake was published in a respected scientific journal and earned him national notoriety.[43]

Louis eventually returned home at about the same time that the British threatened to invade the New Orleans area in late 1814 and early 1815. Upon Louis' reunion with his family, he claimed, without proof, that while living with the Indians he had been made a chief. He made an even more spectacular claim that in his years of wandering he had traveled much of the continent and explored along the Pacific coast where he confided in his brother Doradou that he had discovered the presence of gold. Such a claim nearly four decades before the great forty-niners' gold rush seemed to conservative Doradou to be nothing more than an exaggerated boast by his capricious brother designed to win financial support from the Bringiers. Doradou flatly refused to have anything to do with Louis's dubious scheme to mine the precious metal.⁴⁴

After serving with his younger brother in the American forces at the battle of New Orleans and having his claims of ownership to two tracts of land in "La Fourche" county—which he maintained he had obtained title to during the Spanish regime—rejected by the United States government, Louis became bored with his new sedentary life in Louisiana. Missing the adventurous lifestyle he had enjoyed on the frontier, Louis began to search for a new venture. Having become friends with an old priest who had lived in Mexico for years, Louis was enthralled with the aged prelate's stories of a secret silver mine in Mexico. Not having the resources to take advantage of the news of the mine, Louis went to his brother Doradou for advice and money. Though he was very familiar with Louis' often exaggerated claims of his earlier adventures in the north and west, Doradou, after originally rejecting the idea, eventually relented and gave in to his brother's frequent and fervent requests and provided funding for his brother's new adventure—a search for Mexican treasure.⁴⁵

Louis had a parchment giving directions on how to locate the mine and a ring to present to a woman near the locality of the mine in Mexico. Louis reported later in his life that when he arrived in Mexico he drifted into an area being devastated by a smallpox outbreak. Having a supply of vaccine with him, Louis prevailed upon Indians in the area to let him inoculate them. His generosity and care for the Indians won him their fervent gratitude as demonstrated by their leading him to an ore-rich gold mine. He soon acquired both great wealth and the suspicions of the Mexican government.⁴⁶

Word of Louis' good fortune eventually reached his family back in Louisiana. Any skepticism held by the Bringiers as to the veracity of the stories of their dauntless relative's newfound wealth was removed sometime later when Louis returned to Louisiana for a visit. Seemingly to make a point to those—particularly his younger brother Doradou—who had in the past questioned his adventures, Louis (now called Don Louis) went out of his way to display his wealth. An example of his extravagance occurred during a social at the Hermitage when Don Louis and his male guests entertained themselves by testing their throwing strength and ability by standing on the river bank skimming stacks of Mexican silver coins into the Mississippi River.⁴⁷

Unfortunately for Don Louis when he returned to Mexico, his good luck once again came to a quick end. For reasons that remain unclear, Mexican officials had the extravagant Louisiana Creole imprisoned, confiscated his estate and all of his wealth, and condemned him to death. In spite of his predicament, Don Louis kept his wits. When a priest was sent to his prison cell to prepare him for his execution, he convinced the priest that he was a close relative of Bishop DuBourg in Louisiana and begged the cleric to get into contact with him.[48] He also emphasized that his brother possessed great wealth and would be willing to pay a large ransom for his freedom. Intrigued by this promise of a possible payoff, Mexican officials postponed the execution long enough to get word to and an answer from Louisiana on his promise of a ransom.[49]

Don Louis' pledges to the Mexicans proved to have merit. Both Bishop DuBourg and his brother Doradou responded quickly to word that his life was in danger. The bishop immediately communicated with his counterpart in Mexico, while Don Louis' brother Doradou gathered a large ransom and left posthaste for Mexico. Fortunately, the efforts of the two men proved successful and the one-time millionaire was allowed to leave Mexico with his life and only a few trunks worth of clothing.[50]

On his return to Louisiana, Don Louis worked as a surveyor, including stints as Surveyor for the City of New Orleans and Surveyor General of Louisiana. In his surveyor's role, he played an integral part in the design and development of some of the city's newest neighborhoods of the time, including the growing residential neighborhoods along the Esplanade Ridge. Seizing upon the business opportunities provided by the rapid growth in the city following the end of the War of 1812, the Bringier brothers invested extensively in the real estate and housing industry in New Orleans; working together, they bought and constructed houses on more than one hundred lots in the city and acquired additional lands for their agricultural ventures. Their largest joint business endeavor involved the acquisition and ownership of Houmas Plantation in Ascension Parish. The Houmas estate was so large that its boundaries stretched from the Mississippi River to several miles to the east and north, all the way to the shores of Lake Maurepas. Though meeting with some success, Don Louis' new business ventures in Louisiana never provided him with the triumph or fortune equal to those he claimed to have acquired and lost during his ill-fated Mexican adventure.[51]

However, his new business activities provided him with a more than comfortable lifestyle. The construction of his New Orleans' residence on the city's fashionable Esplanade Ridge demonstrated the extent of his financial well-being. Built on a lot bounded by Esplanade, Prieur,

Kerlerec, and Roman streets and surrounded by a beautiful garden, the house was constructed in an architectural style reminiscent of the structures he had known in Mexico. The imposing house included large pillars and arches and amazingly encompassed on its roof a large awning-covered fishpond where Don Louis spent much of his leisure time fishing.[52]

His death at age seventy-six on October 29, 1860, did not mark the final chapter for the Bringiers of Don Louis' life and his Mexican exploits. Years later, a young Mexican male arrived in New Orleans and approached the male members of the Bringier family and informed them that he was the son of Don Louis. The information he provided of the love affair between Louis and his mother was such that they were convinced of the veracity of his story. The family provided the young man with a purse which he used to open a barber shop that male members of family patronized. In line with the uncompromising social customs of the time, the young Mexican—whose name was not recorded in the records kept by the family—was never intentionally introduced to any of the female family members and was never invited to Melpomene, the Bringier family's imposing city home.[53]

Don Louis' Mexican son prospered as a barber and never forgot the generosity of the Bringiers. Years later when Don Louis' daughters were reduced to poverty, the barber privately approached his half-sisters and provided them with financial assistance. The daughters Letitia and Louise responded to his generosity by inviting him to their home. News of this social transgression greatly disturbed and scandalized the rest of the family. Learning of the family's disapproval, Letitia dismissed their outrage with the comment that she owed profound gratitude to her Mexican brother who came to her assistance at a time of need and that she owed nothing to her wealthy relatives who had snubbed her and ignored her plight. She concluded her response with the terse observations that "I think a great deal more of him than of the proud legitimates of Melpomene."[54]

Letitia's pithy response is demonstrative of the strained relations that developed between Don Louis' family and the rest of the Bringiers. The schism developed as a result of Louis' marriage in October 1831, as he neared age fifty, to a beautiful young—yet poor—woman. Although Doradou and Louis' relationship remained close, the younger brother's family disapproved of the marriage between Louis and the beautiful Marie Josephine Hermione Guignan.[55] Relations between the families of the two Bringier brothers remained cordial until Doradou's death in 1847. From then on contact between the two families became rare and formal. After Louis' passing and for years hence, Doradou's widow Aglae once a year made a formal visit to Louis' family. Unfortunately, the widow Hermione suffered from a mental illness that grew progressively worse as the years passed. The adult members of the Bringier family were concerned about her condition and even made an offer, rejected by her children, to cover the cost of having her institutionalized. For many years, Hermione's mental condition served as a center of interest, discussion, and even sport for the Bringier children.[56]

Although the lives of Marius Pons' other children did not match the level of adventure experienced by their brother Louis, they did share their brother and father's business skills and adroitness. Somewhat surprisingly, given the social order of the time, which ordained well defined gender roles that limited the business activities of women, Marius Pons' eldest daughter Francoise demonstrated a remarkable degree of business acumen and independence as the manager of her own plantation. Born on March 9, 1786, at the family's White Hall estate, Francoise and her younger sister Elizabeth preferred to be known by the more fashionable English equivalents of their French names— Fanny and "Betzy," this in spite of the fact that the family nearly always spoke using their father's native tongue of French.[57]

Strong willed and determined, Fanny inherited her father's drive and fiscal prowess. Her skills in this area became evident to all when she assumed daily management control of the Bocage plantation where she and her husband Louis Christophe Colomb lived. Given to Fanny and her husband by her father Marius Pons at the time of the couple's wedding, their Bocage plantation was located on the east bank of the Mississippi River in Ascension Parish adjacent to the Hermitage plantation of her brother Doradou.[58] Originally part of the family's large Houmas lands acquisition, the couple named the plantation for a small shady wood or retreat.[59]

Partly through her love for business activities and partly by necessity caused by her husband's complete lack of interest or ability in running their estate, Fanny assumed, which at the time was an unusual role for a southern lady, the job of plantation manager. Born in 1770 in Paris, her husband Christophe, took part in the French Revolution and fled France for political reasons. Disguised as a cook on a vessel sailing to America, Christophe made his way to the French colony of San Domingo. Unfortunately, his stay on the Caribbean island did not last long. Shortly after his arrival, the island colony erupted with a bloody slave rebellion which forced him and many of the other whites on the island to flee for their lives. Like many trying to escape the conflict, Christophe made his way to the French community at New Orleans to begin life anew. Upon arriving in the Crescent City, he supported himself as an artist.[60]

It was as a new resident in Louisiana that Christophe became friends with the young and adventurous Louis Bringier. After the wild escapade, discussed previously, in which the young son of Marius Pons Bringier lost a large sum of his father's money while on a business trip to New Orleans, the older Christophe befriended the troubled youth and agreed to accompany him back to his father's estate to help explain what had happened. The elder Bringier was so impressed with Christophe's maturity and presence that Marius Pons not only forgave his son for his irresponsible behavior, he invited the Frenchman to stay as a guest at his White Hall manor house. Not long afterwards, Marius Pons bestowed the hand of Fanny, his eldest daughter, on Christophe and shortly thereafter on January 26, 1801, the groom, at approximately age thirty-one, married his young sixteen-year-old bride.[61]

No doubt to some bewilderment on the part of his young bride's family, Christophe soon let it be known that he was an artist, a poet, and a musician and not a plantation manager. He possessed no interest—and seemingly little ability—for the tough and demanding job of running a plantation. On the other hand, his young bride loved the challenge of assuming the role of operating the estate and was often seen riding about the plantation on horseback inspecting the operations of the estate and giving directions to her workers, much as her male counterparts did on their neighboring plantations.[62]

With his wife Fanny handling the estate's business matters, Christophe was free to indulge his passions of art, poetry, and music. Among his most popular activities was being rowed by slaves from the estate on the Mississippi River in a special ornate rowboat that he had personally designed, while he reclined on cushions, strummed his guitar, and crooned songs of love. In addition to the singing of his love songs, nearly every day he would have workers row the mile or two to Donaldsonville on the other side of the river where he spent time with friends—or anyone else who would give him the time—conversing and playing cards. His agreeable manner, liberal education, and stories about the French Revolution and Napoleon made him one of the most popular visitors to the town of Donaldsonville.[63]

In spite of the unusual nature of the relationship, the marriage of

Fanny and Christophe was a happy one. Each partner was content with their role in the relationship. Fanny considered her husband to be brilliant and often noted how proud she was of his brilliance. The happy union of Fanny and Christophe lasted for over two-and-a-half decades until her death on May 10, 1827. Unfortunately, following Fanny's demise, Christophe's relationship with several members of the Bringier family grew increasingly strained. After the death of his beloved wife, Christophe traveled to the north where he lived for a period of time. While there, he met and married Helen Belton, a resident of Maryland and a member of the Perry family of America's two naval heroes of the same name. After the marriage, Christophe and his new wife returned to Louisiana to live at Bocage.[64] Regretfully, the new Mrs. Colomb simply did not get along with many members of the Bringier family. Most notably, she frequently quarreled with Christophe and Fanny's children over the operation of the Bocage estate. The bickering continued until Christophe's death on March 9, 1832.[65]

Christophe Colomb was not the only refugee of the San Domingo slave revolt to become a member of the Bringier family. Marius Pons' second daughter Louise Elizabeth was born at White Hall on April 21, 1788. She preferred to be called by her English nickname of Betsy; however, the family members put a French twist on their enunciation to come up with the pronunciation of Bet-zee. As was the case with her older sister Fanny, Betsy was betrothed by her father to an older man. At the age of fourteen, her father pledged the hand of the young Betsy to thirty-nine-year-old Augustin Dominique Tureaud.[66]

A native of France, Augustin, like his future brother-in-law Christophe, was a refugee of the San Domingo slave rebellion. Escaping to the United States, he spent time in Baltimore, where he became a U.S. citizen. Moving to Louisiana shortly before the American acquisition of the territory in 1803, he became an acquaintance of Marius Pons. Tureaud, a merchant in New Orleans, was invited to White Hall to inspect the cotton crop that Marius Pons offered to barter to Tureaud's commercial house in exchange for imported items. Poor weather conditions required the merchant to extend his stay at the plantation for a fortnight. During his stay, Tureaud and Marius Pons established a relationship, with Tureaud being treated by the planter with what he called "friendly confidentiality." Though Tureaud was in his late thirties, he became acquainted during his stay at the Bringier house with the planter's fourteen-year-old daughter, "little Betsy," whom he depicted as "without being beautiful, was more good looking than otherwise."[67]

Before Tureaud's visit at White Hall ended, Marius Pons approached the ambitious merchant, who was a widower with several teenage children living in his native France, with an offer to establish a business and personal partnership. Noting his need for someone to represent his cotton interest in New Orleans, Marius Pons offered to establish a firm under Tureaud's direction. The firm would be funded by four $8,000 investments coming from the planter, a son, a son-in-law, and Tureaud himself. Marius Pons was so serious in making the deal that even when Tureaud admitted that he did not have the funds for his share, the planter stated that factor would not be a fatal obstacle to the deal. However, according to Tureaud, the Creole planter noted he had one additional consideration and stated the following:

> ... in wishing to associate myself with you in interests, I wish it to be an association of the closest nature. I sincerely desire that my daughter Betsy should be the immediate means of cementing our union. See her. Speak to her of my intentions, and if she shows no disinclination to unite herself to you, everything can be arranged in a few days. I must observe, however, that I do not want to cross the wishes of my daughter for any consideration whatsoever. So this I must add: if the marriage does not take place, the partnership will not be formed.[68]

After several days of the couple getting to know each other and discussion on the possibility of a marriage between the two, Marius Pons called the two together and asked his daughter if she agreed to marry the merchant. According to Tureaud, "[she] placed herself in my arms, and the kiss she gave me made me understand that her mouth was the interpreter of her heart." However, Betsy's fervor for the marriage, though positive, appears to have been somewhat less enthusiastic than Tureaud's recollections. In at least one letter written at the time of her marriage to the two-and-a-half-times her senior groom, she referred to him as the "gray-headed old man." Both the business nature of the nuptials and age differential were common for the time and did not appear to hinder the festive nature of Betsy's wedding, which like those of most of her siblings was held in grand style at White Hall on May 24, 1803. Although the marriage was more of the result of a business deal than a love affair of the couple, the relationship between the pair and the Bringier and Tureaud families remained extremely close for decades.[69]

Tureaud quickly established himself as one of the prominent members of New Orleans society. Due to his prominence, the French Prefect chose Tureaud to serve as a member of the new city council, which was created to replace the Spanish Cabildo when ownership of the colony of Louisiana was transferred from Spain to France and shortly thereafter to the United States in 1803. Tureaud continued to hold positions of responsibility even after the Americans took control of Louisiana. In 1805, the American governor William C. C. Claiborne commissioned Tureaud as a captain in the new militia of the Territory of Orleans and in 1811 as the Justice of the Peace for his home parish of St. James. He and Betsy resided at their Union plantation until his death on April 16, 1826.[70]

Marius Pons' second youngest child Francoise Laure Bringier was born at White Hall in 1792. She was known throughout her life by her middle name Laure. Like her two older sisters, she married a native of France. In a grand ceremony at her father's plantation on May 12, 1810, she married Noel Auguste Baron, Jr., of New Orleans. A native of Normandy, France, Auguste was one of the region's most successful commission merchants. In the same year of the marriage, he formed a business partnership in New Orleans with Pierre Francois DuBourg. The new firm of DuBourg & Baron soon became the agents of the region's most prosperous planters. However, unlike her sisters, Laure and her husband did not choose to live on a country estate. Until his death on June 10, 1833, Noel and Laure resided in New Orleans in a house on St. Charles Avenue not far from Canal Street and near the site that would later become the location of the city's famous St. Charles Hotel. Unfortunately for Laure, the death of Noel in 1833 left her with limited resources to support their seven children. Her situation became so difficult that in 1845 her brother Doradou and her son Jules Baron purchased a house on Edward Street in New Orleans for her and her younger children. All of Noel and Laure's children, except for the youngest child, lived to adulthood and married. Their third child, Francoise Adelaide, married Francis Richard Lubbock in 1835. A native of South Carolina, Lubbock moved with his young bride to Texas where he opened a general store and entered Democratic Party politics, which led to him serving as the state's Confederate governor from 1861 to 1863.[71]

Marius Pons' sixth and final child Melanie Elizabeth was born on August 16, 1793. The young girl was betrothed to William Simpson, a native of Savannah, Georgia, and a friend and business associate of her father. Simpson served as an agent for both Marius Pons and his son Doradou. He also speculated in real estate investments, including joining with John Watkins in purchasing twenty-thousand arpents of land along the Amite River in the St. Helena portion of the Baton Rouge District. After their marriage, the young couple lived in New Orleans where Simpson continued his investments in real estate. Among his holdings was the square of ground near Laure's house on St. Charles Avenue that would become the future site of the popular St. Charles Hotel. Before any children were born, William died at age thirty-nine in January 1813. Widowed at age twenty, Melanie moved back to her parent's White Hall plantation. Inheriting her husband's business holdings, Melanie grew close to one of her husband's clerks, James Fisher Wilson. Eventually the dealings between Wilson and Melanie evolved from being a business association to a personal relationship. The couple married in June 1816 and eventually had three children.[72]

Though all of Marius Pons' children lived to adulthood and enjoyed comfortable lifestyles, the child whose life most closely resembled that of their father was his youngest son, Michel Doradou. For unlike his older brother Louis, Doradou possessed little of the wanderlust that drove his brother's thirst for adventure, fame, and fortune. Doradou instead loved the stable and cultured life enjoyed by the planters in southern Louisiana. Thus it fell to him, more than any of his siblings, to carry on the presence and traditions established by Marius Pons at White Hall estate.

Chapter 3

Doradou and Aglae

All of Francoise Durand and Marius Pons Bringier's six children born in Louisiana lived to adulthood and gained community standing. None, however, had a more lasting influence upon the Bringier family and the development of the plantation culture of Louisiana's river region than their second son, Michel Doradou Bringier. Born at sea on December 6, 1789, while his parents were on a trip to the West Indies to obtain items for the construction of the new house at their White Hall estate,[73] the new family member was likely named for a family associate, Michel Cantrelle, who served for many years as a judge in St. James Parish in Louisiana. His middle name was chosen to honor the family of his father's grandmother—Douradou—from whom the Bringiers inherited a great deal of property. Later in life when he chose to go by his middle name, he modified the spelling to Doradou. As an adult with a family of his own, he was commonly called "Bon Papa" by his wife and children.[74]

Doradou married in his early twenties. A common practice at the time, his marriage into one of Louisiana's most prominent families was arranged by his and his bride's relatives. Pierre DuBourg was one of the leading citizens of New Orleans during the colonial period.[75] As a onetime Collector of the Port of New Orleans, a former officer in the citizen militia for the Territory of Orleans, and one of the leading commercial agents in New Orleans, DuBourg was acquainted with most of the leading families of the region. Included among his longtime acquaintances and customers was the owner of White Hall plantation, Marius Pons Bringier.[76]

Pierre DuBourg's story was not unlike that of Marius Pons. Brought up in an atmosphere of wealth and influence in Europe, as a young man he moved to the San Domingo colony. Located on the island of Hispaniola, it was the richest and most populated of France's colonies in the New World. The revolt of the island's slaves from 1791 to 1804 led many of the island's French residents to flee their homes for safer lands. DuBourg joined the exodus of Europeans from the island and eventually made his way to New Orleans by way of British Jamaica and Baltimore, Maryland. Gaining prominence and positions of some authority in New

Orleans at the time the city was undergoing its transformation into an American possession, he had the ability to get along well with both the newly arrived Americans and their more conservative *ancienne* neighbors of French and Spanish heritage.[77] This was no easy task. Because of the many cultural and religious differences of the two groups, relations between the Louisianians of European and those of American heritages remained strained for years following the acquisition of Louisiana by the United States. His ability to relate to and associate with both groups was evidenced by the fact that of his five daughters, three married Americans and two married members of the *ancienne* culture. Furthermore, he was also among the organizers and founders in 1812 of the Grand Lodge of Masons of Louisiana and was subsequently chosen as the First Grand Master of the organization. Although he spoke English perfectly and remained close to the Americans in Louisiana, DuBourg never personally abandoned his French heritage. Throughout his life, in his home and later even in those of his children, the family communicated almost exclusively in DuBourg's native language of French.[78]

As one of the region's leading commercial agents, DuBourg and Marius Pons Bringier often did business together. Over time, their business partnership evolved into a personal friendship. As the relationship between the DuBourg and Bringier families grew closer, the two patriarchs decided to form a permanent alliance between their families through the marriage of Doradou and Pierre's daughter Elizabeth Aglae DuBourg.[79]

Born in Kingston, Jamaica, on January 4, 1798, Aglae at the age of nine was taken to Baltimore, Maryland, to live under the auspices of her uncle and godfather, the Catholic prelate L'Abbé William DuBourg.[80] According to family tradition, she first met her future husband in New Orleans while traveling to her new home in Baltimore. This first meeting between seventeen-year-old Doradou and Aglae made a great impression upon the young Bringier. Later in his life, he often proclaimed that upon meeting her he felt that she was the most beautiful child that he had ever seen and was supposedly struck with love at first sight. Over the next five years, the two families grew closer with Aglae's father and uncle eventually entering into an agreement with Marius Pons that the two families should be joined by her marriage to Doradou. A willing participant in the matchmaking effort of the family elders, Doradou immediately set out for Baltimore to claim his young bride upon learning of the marriage pact. Only fourteen at the time, she was eight years younger than her betrothed.[81]

Family accounts reveal that her uncle, the Abbé DuBourg, was the moving force behind the betrothal. His family lost its fortune during the French Revolution and San Domingo slave revolt, and he no doubt saw the potential for a profitable alliance with the wealthy Bringiers. Though the betrothal of Aglae to Doradou was the first match orchestrated by the prelate, it was not the last. Building upon the success of his effort with the Bringiers, the Abbé DuBourg played an ever increasing role in conducting both the educational and matrimonial affairs—including the picking of proper husbands for them—of the young females in his family.[82]

During the intervening years between the time the young couple met each other in New Orleans and the time when arrangements for the marriage ceremony were finalized, young Aglae remained in Maryland. While residing there, she attended the school of Madame Lacombe for a year before transferring to the school of Mother Elizabeth Seton where she remained until the time of her marriage. Her father and uncle saw to it that young Aglae received the rigorous training in domestic etiquette young ladies of the old regime were expected to master. Upon learning that Aglae was to be one of Seton's first students, her father wrote his brother the prelate giving his support to her attending the new school and noted that:

> Bonne and I are perfectly tranquil about your care for our dear Aglae and do not doubt that under your eyes and with your good advice she will become very interesting. May she continue to study with attention to French and English and all the skills necessary to her sex. . . . The new arrangement you have made can only content us since we rely entirely on your good judgement and tender affection.[83]

Commonly called "fireside etiquette" at the time, the training of a young girl included the basics in reading, writing, arithmetic, and moral training. Girls also received courses in grammar and composition and were given special instruction in the form and style of letter writing. Reading popular novels of the time was not recommended for young ladies; instead, emphasis was placed on the study of Scriptures. Other common topics of study for girls included ancient and modern geography, languages, natural history, and astronomy. In the area of etiquette, young girls were usually prompted along in their training with stiff doses of discipline and rigid rules. For example, Aglae and her sisters were expected never to lean their elbows on the arm of a chair or let their back touch the chair. Years later her children marveled that even in her old age and in the privacy of her own room, she adhered to her childhood training and never allowed her back to rest against a chair. A woman of remarkable beauty in her youth, her comely appearance and the courtly manners learned in her youth were retained until her death. A person of quiet and reserved manners, her great-grandson and Bringier family chronicler, Trist Wood, lauded in his recollections of her "uncommon degree [of] both sweetness of disposition and dignity of manner."[84]

The couple's wedding took place in Baltimore, on June 17, 1812, a day before the United States declared war against Great Britain at the start of the War of 1812. The bride's uncle and godfather, the Abbé, represented the DuBourg family. Surprisingly, her father did not attend the ceremony, considering the social importance given to weddings during the era by the social elite. Though there is no evidence that he opposed the marriage, since it could not have taken place without his blessing, Aglae's father did not travel to Maryland to participate in any of the events associated with the special occasion. It appears that his involvement in the inauguration of the Grand Lodge of Masons of Louisiana and his selection in New Orleans as its First Grand Master, which occurred only three days after the wedding in Baltimore, unavoidedly and surprisingly took priority over his daughter's wedding.[85]

As was common at the time when two families of substance were involved in the arranging of a marriage, the Bringiers and DuBourg formalized their negotiations through a prenuptial agreement. On the same day of the wedding, Doradou—representing himself—and Louis DuBourg—Aglae's uncle, representing the fourteen-year-old bride—met at a notary public in Baltimore and signed a formal prenuptial agreement. Noting that Doradou's wealth amounted to $10,000, his Hermitage estate of approximately 800 acres and a cadre of thirty-two slaves and that Aglae's fortune totaled $1,000 and consisted mostly of linens, jewelry, clothing, and other goods, the document called for a partnership of community property and a $1,000 gift from Doradou to Aglae in case she survived him.[86]

At the time of the marriage, Baltimore had a large settlement of French residents who had fled to the city from the revolution in France and the slave revolt in San Domingo. The wedding caused considerable interest in the Maryland city because of the prominence of Aglae's uncle L'Abbé DuBourg—who had served as President of Georgetown

University in nearby Washington, D.C., and had established l'Academie DuBourg in Baltimore—and the groom's prominence as the son of one of Louisiana's leading French families. Additionally, the extreme youthfulness of the bride intrigued many of the DuBourg family's American friends and acquaintances in the city. Although Southern females on average married at an earlier age than their Northern sisters, marriage at the age of fourteen was relatively unusual. A study by historian Catherine Clinton revealed that only .5 percent of women in the South at the time of Aglae and Doradou's wedding married at age fourteen. Clinton's study demonstrates that both the bride and groom were somewhat younger than the Southern norm of the day, with the median age being twenty for women and twenty-eight for men at the time of their marriage. Paul Lachance's marriage study of antebellum New Orleans found that as in the case of the Bringiers, a large difference in ages between spouses was common at the time and that 58 percent of brides in the city were minors on their wedding day. His analysis of the marriage participants' ages shows that the average age in antebellum New Orleans of brides and grooms at the time of their marriage was twenty-one for brides and thirty-two for grooms—ages somewhat higher than those of the Bringier couple. One disconcerted American friend gave a beautiful baby doll as a wedding gift to the young bride. Somewhat flustered by the gift, Aglae good naturedly commented upon unwrapping the present, "I do not know whether this is intended for me, or for my first baby!"[87]

Immediately following the ceremony, the young couple, along with the bride's uncle who had recently been appointed the new Apostolic Administrator of the Catholic Diocese of New Orleans, sailed for the Crescent City. Faced with contrary winds, the ship was forced to put in temporarily at Norfolk, Virginia. After a couple of days of delay, the vessel continued on to Havana for another overlay of several days and thence onto New Orleans. Although there was some fear of interception by British warships, there was relatively little chance of this occurring at that early stage of the war.[88] Upon returning to Louisiana, Doradou set to work constructing a stately home for his new bride. For the location of their new home, he chose a site on the east bank of the Mississippi River in Ascension Parish, several miles away from the family's White Hall homestead. Doradou acquired the property in a business arrangement with his father several years earlier in 1806 when he was only seventeen years old—approximately the same time that the Bringiers and DuBourgs agreed to have the young couple marry. However, the young Bringier did not obtain full and sole ownership of the property until 1812.⁸⁹

Possessing a healthy, though not plush, personal fortune of approximately $11,000 and thirty-two slaves, young Doradou did not have the resources on hand needed to build a home that would meet the standard he wished to provide his beautiful young bride. According to family tradition, the elder Bringier was so pleased with his son's marriage and new bride, that he presented the young couple with a wedding gift—the means required to construct an imposing home and estate for them to live and have a family.⁹⁰

Although work on the couple's new home started soon after their marriage, the ongoing war with the British hampered the construction. Wartime shortages and the presence of blockading British warships in the Gulf of Mexico hurt the region's economy and made acquiring some supplies and materials for the house a difficult task. Fortunately with most materials needed for the construction of the structure available locally, work on the house continued to progress. The only interruption in the project occurred in late 1814 when an invasion by British forces threatened the region. On November 18 of that year, Doradou enlisted in the military company of Capt. J. Chauveau. In late December and early January his service took him to the Chalmette battlefield a short distance down river from New Orleans where Gen. Andrew Jackson led his American force in a successful defense of the city against Gen. Edward Pakenham's British army. During the heroic struggle, Doradou served as a volunteer aide-de-camp to General Jackson. He remained on duty until March 14, 1815, when word reached New Orleans that the Treaty of Ghent had been signed ending the war and the threat of another British attack.⁹¹

During the months that he was on active duty, Doradou worked closely with General Jackson. The young Creole soldier became a great admirer of the American hero and remained loyal to Jackson and his political party. Details are scarce as to how the young Creole officer and his family became close to the crusty American general. Regardless of how the association began, the Bringiers and Jacksons remained friends throughout the life of the future American president. Young Doradou was so taken with Jackson that on his return home after serving with the general he decided to name his and Aglae's new home "l' Hermitage" in honor of the general's home in Tennessee. However, over time as both the Bringiers and the local community became more Americanized, the estate was increasingly referred to as "The Hermitage." By the end of the antebellum period, the estate was almost exclusively referred to by its American name.⁹²

Over the years, the Bringiers' loyalty and friendship with Jackson remained strong. When the then ex-president made his last visit to New Orleans in 1840, Jackson insisted on taking time from his busy schedule to have the captain of the steamboat *Sultana*, on which he and several members of the Bringier family were traveling, stop at the home of his former aide-de-camp for a brief visit. To memorialize his visit to the Hermitage and to honor his hosts for their hospitality and long friendship, the former president cut a lock of his hair and presented it to the family.⁹³

Unlike his father and flamboyant brother, Don Louis, who both enjoyed being in the public eye, Doradou, except for his short stint of militia service, never aspired to public office or attention. Though wealthy and influential with a large number of friends, throughout his life he concentrated his time and attention on his family, his real estate holdings, and his plantation with all of the various business dealings associated with the operation of a large staple-crop agricultural unit.⁹⁴ Though he avoided public notoriety, Doradou worked hard to improve his status. His efforts paid off and he quickly obtained great financial success, amassing an estate of approximately $1,700,000. Becoming the wealthiest member of the Bringier family and one of the wealthiest individuals in the entire state of Louisiana, he was also blessed with great organization and business skills. Possessing exceptional common sense, he was known for his shrewdness and ability to identify and remain focused on the issue at hand. Scrupulously honest in his business and personal dealings, he earned the respect of both proponents and opponents.⁹⁵

However, like others in the Bringier family, he sometimes possessed a violent temper. Though slow to anger, once he did he often reached a point of unreasonable outrage. An example of his temper, and its cost to him and his family, occurred when he was in negotiations to purchase a large tract of land above Canal Street in New Orleans. After all arrangements had been agreed upon, the parties met for the document signing. When a technicality developed over what medium of payment—whether gold or silver—would be used for the property purchase, Doradou could not understand the sellers' objections. He had offered payment at a full and stable rate. In spite of his assurance of sound remuneration, the property owners continued to question the issue. Doradou arose in a rage, grabbed the two vendors by the nape of their necks, bumped their heads together, and declared that the deal was off! Unfortunately, for Doradou and his descendants, the real estate deal could have brought the family considerable additional wealth. The land in question made up much of the area that became the Faubourg St. Marie. Later owned and developed by Jean Gravier, the area became the center of the city's American sector and contained some of the most valuable property in nineteenth-century New Orleans.[96]

As was popular among the residents of the region, Doradou loved taking part in the many outdoor activities, which proliferated in the woods and waterways in the vicinity of the Hermitage. The area's many bayous, rivers, and lakes abounded with ducks and other wildlife. Wild deer, bear, rabbits, squirrels, and quail were available in great numbers in the fields, forests, and swamps on or near Doradou's properties. Both day and weeklong hunting and fishing trips were common events enjoyed by the Bringiers and other families of the region. Doradou particularly loved hunting. He owned a special hunting piece, which he alone could handle. This special gun was designed to use a larger charge of powder than a traditional hunting gun. The report of the weapon was so loud that it sounded like "a young cannon." Doradou was so proficient with his special gun that when hunting with others he would often wait until all other members of the hunting party had fired at the game. If they all had missed, he would level his weapon and with unerring accuracy bring down the prey. When his son Amedee recounted to his nephew Trist Wood his youthful remembrances, he related that as a boy hunting with his father he could not remember a time when his father had failed to hit the object for which he was aiming.[97]

In addition to being a dead shot, Doradou also possessed great physical strength and vigor. Up until the time of his death at age fifty-seven, family members claimed that he could physically overpower all of the youth about him in any athletic sport of their choosing. In spite of his physical prowess and competitive disposition, with the exception of fatherly play with his young children, there was seldom a need for Doradou to demonstrate his impressive strength with family members. As was customary among most plantation families in the antebellum South, Doradou left the responsibility for maintaining discipline among the children to their mother. Although she accepted her role as family disciplinarian, she did not relish the job and complained to her husband, for she felt it put a strain on the relationship between her and the younger children in the family.[98]

Doradou and Aglae had a total of ten children; three sons and six daughters lived to adulthood.[99] In addition to her role of family disciplinarian, Aglae was responsible for the children's education. Attitudes in Louisiana on education were not unlike those in the rest of the South at the time. During the colonial period, with the exception of a few Catholic religious orders, there was little interest placed on education. However, with the onset of the American regime in Louisiana, increased attention was given to the issue. The American Revolution and its accompanying ideals of republicanism led to a common belief in the nation that education was a necessary ingredient in a democratic society. During the post-revolutionary era, Southern upper-class families increasingly showed a concern for obtaining adequate educational opportunities for their children. However, what was considered adequate varied from region to region, family to family and, even more frequently, between brother and sister. Generally, greater attention was given to educating sons so that someday they could assume their proper place in society as gentlemen leaders and managers of their own businesses and estates. The schooling of males often lasted years longer than that of their sisters and, as was the case with the Bringiers, frequently included attendance in out-of-state institutions and even schooling in Europe. On the other hand, schooling for females during the period was less rigorous. Whereas males often received wide-ranging training in the classics, philosophy, history, and mathematics, females were provided with a more "polite" education centered on reading and writing. Emphasis was placed upon grammar and composition, with special attention given to the skill of letter writing and the study of the Scriptures. To aid upper-class Southern girls in managing their household accounts and in understanding plantation ledgers, young ladies studied the fundamentals of mathematics. As was common practice, the younger Bringier children received most of their formal education from tutors, who either lived at or visited the planter's home to teach the estate's children. Among the tutors employed by the Bringiers were two sisters, Estelle and Fanny Baconais. However, when in the city, the Bringier children attended nearby schools, including the Female Seminary, which was located at the present site of Lee Circle on St. Charles Avenue. The school was located only a few blocks away from the Bringiers' Melpomene home and was easily reached by walking along a series of raised walks and open fields.[100]

As the children matured and circumstances permitted, the Bringier children turned to more distant venues for their education. The schooling of each son was individually planned with no evidence of any two brothers attending the same institution. Young M. S. Bringier attended school at Mr. Islley's in Donaldsonville and later studied under the L'Abbé Anduse in Iberville Parish. As M. S. matured, his great-uncle, Bishop DuBourg, played an active role in planning his education. The bishop arranged to send him to Bardstown, Kentucky, for additional schooling and later to Europe to travel and to complete his studies. Similarly, Amedee Bringier attended Jefferson College at Convent in St. James Parish, located not very far from the family's home at the Hermitage. As a young man, he later traveled to Charlottesville where he entered the University of Virginia. The youngest of Doradou and Aglae's boys, Martin Doradou—called by the family "Doradou," "Dadou," and "Junior"[101]—was also sent east for his formal education. He obtained most of his education at Mt. St. Mary's near Baltimore, an institution founded with the help of his great-uncle, Bishop DuBourg. As he grew older, the Bringier's youngest son switched his schooling to an institution at Bloomfield, Virginia, where he remained enrolled as late as 1858.[102]

The family also traveled for business and entertainment reasons. Doradou's business ventures frequently took him on extended trips away from his family. Family members were fortunate that their financial resources permitted them to also enjoy the pleasures of occasional vacations. The Bringiers often made extended trips to other parts of the country to enjoy the change in scenery sometimes taking their vacations with other relatives—most notably with their daughter Nanine, her husband Duncan Kenner, and their children. Among their favorite destinations were summers spent at Newport, Rhode Island, and stays at the popular White Sulphur Springs in Virginia, where there were comfortable accommodations in large, well-accommodated hotels and cottages. Closer to home, the family spent most of their summers in or near Louisiana at various places across Lake Pontchartrain from the city—a commonplace practice among the planters and wealthy residents of the Crescent City. Out-of-town travel was popular because of the poor health conditions that usually prevailed during the summer months in subtropical New Orleans and because of the slack in planting activities on sugar plantations at that time of the year. Like many others, the Bringiers developed a preference for spending their summers along the Gulf Coast, including stays at Bay St. Louis, Biloxi, and Pass Christian. The gulf breeze, high ground, and pine forest offered a refreshing change from the uncomfortable weather conditions found back home. The convenience of having a nearby escape led Aglae Bringier in

April 1854 to purchase a house at Bay St. Louis from George Fosdick. Thereafter, the new house "across the lake" became the popular center of summer activities for the many members of the extended Bringier family. Shortly after Aglae bought the house on the bay, her son-in-law Martin Gordon, Jr., in a letter to Benjamin Tureaud, who was already on the coast, expressed his unhappiness at not being able to join the family on the coast when he wrote, "Doctor Campbell and his gang take their departure today for the Bay of St. Louis; on Friday next, Kenner and his troupe will go over; you will soon have a devil of a Crowd [sic] at the Bay." He ended his letter with the sigh, "I wish to God I could leave the City."[103]

In addition to playing together during their vacations, the Bringiers also worshipped and prayed together. Practically all planters believed in an omnipotent Supreme Being and that He sanctioned the South's social order. Across the South, Protestant sects, especially Episcopalianism and Presbyterianism, dominated the denominational preferences among planter families. However, in South Louisiana, Roman Catholicism prevailed among both the population as a whole and the sugar planters of the river region.[104] The Bringiers' dedication to their Catholic faith remained strong across the three generations of the family discussed in this work. Matters of faith appear to have been openly discussed among the family as demonstrated by M. S. Bringier's comments in a letter to his mother in 1849 when he noted "I've been quite happy to learn that Rosella [his sister] is thinking about her religious duties." He added a comment that encapsulated his dedication to his faith and the depth of his providential view of life by noting that:

> . . . we have a real need for religion and it is incontestable that the Catholic religious beliefs ease the moments of pain and sorrow that encompass, as it were, so much of human existence. Happy is he whose haughty pride can adapt itself to religious beliefs, for even in his misfortunes religion can provide him with a consolation, perhaps the greatest, that is the idea and the thought of a reward in after life.[105]

Although existing family records and recollections reveal a loving and close-knit family, the documents also establish that the Bringier household functioned with a polite formality. Doradou, Aglae, and the children each had specific areas of responsibility and roles they filled in the family. No clearer demonstration of the formal nature of the family dynamics can be found than in the circumstances associated with Doradou's death. Though a person of robust health and strength throughout most of his life, Doradou's health eventually failed when he was stricken with mouth and throat cancer. Little is known of this final period of his life because he chose not to discuss his medical problem and its potential consequences with his family members. Even when he left home for the last time to travel to Memphis for treatment, he traveled alone. Only when the surgery failed and the family was informed of Doradou's weakening condition did anyone in the family travel to be with him. Only his sons M. S. and Amedee were able to reach Memphis in time. Regretfully, the Bringier brothers, traveling separately, arrived only a short time before their father breathed his last. Doradou died in his sleep at approximately one o'clock in the morning on March 13, 1847, at age fifty-seven. Unfortunately, because of either the advanced nature of the disease or because of the opium he was given to relieve the pain, he was barely able to recognize his sons during their final meeting.[106]

A steamboat returned Doradou's body to Louisiana where he became the first of many family members laid to rest in the multi-tiered and vaulted family tomb in the Catholic Cemetery in Donaldsonville. Known locally as "Le Monument," or the Bringier mausoleum, Doradou, as he had with the design of his home, spared no expense in the construction of his family's final resting place. Built largely with imported Italian marble, the grand edifice cost Doradou $20,000—an impressive sum for the time. Completed under Doradou's supervision sometime before his death, the grand tomb was designed large enough to serve as the final resting place for many members of the Bringier family.[107]

At the time of his death, Doradou's estate was said to total approximately $1,700,000, making him one of the wealthiest men in the state. Nevertheless, his passing placed a considerable burden—both personal and financial—upon his family, since no will was ever found. The question of the will puzzled family and friends for some time after his death. It surprised many that he left no will especially since he was a successful business man who was known for his formalities and methodical habits and who suffered from an illness that forewarned of possible death. Even more confusing was the fact that one servant, Dedé, always asserted that she had been present at a time when Doradou signed a will. Family members searched for the document in the months following his death. His daughter Mrs. Louise Gordon contended that the will had been burned at the Hermitage. No explanation as to why such an act occurred or why she would make such a claim has ever been found among surviving family papers. Family members continued to debate Doradou's missing will for years. For example, Doradou's grandson, Dr. Trist Bringier of Tezcuco plantation, acquired the writing desk used by his grandfather at the Hermitage. On obtaining the furniture, Trist had it taken apart piece-by-piece hoping that there existed a secret hiding place containing the long-sought-after document.[108]

Dying intestate caused his loved ones considerable difficulty in handling his succession. However, not all of Doradou's holdings were involved in the family's succession debate. Several years prior to his death, he transferred management of some of his land holdings to his son M. S.[109] In 1839, Doradou had acquired a massive tract of land known as the great Houmas grant. The property extended from the Mississippi River back several miles through fertile alluvial soils and impassable cypress swamps to the shore of Lake Maurepas. On this land tract in Ascension Parish, Doradou established the family's Houmas and Brulé estates. "The Houmas" or "Les Houmas" estate took its name from the Indian tribe of the same name and should not be confused with the more famous Houmas House, once owned by Gen. Wade Hampton and also located in Ascension Parish. The Brulé plantation was located to the rear and adjourned the Houmas property. Though some claim that the estate received its name from the old practice of burning down trees in order to clear the land for cultivation, family historian and genealogist Trist Wood contends in his notes on the family that like "The Houmas" name, Brulé was also taken from a local Indian tribe.[110]

Continuing the practice started by his father with him, Doradou entrusted the management of his Houmas and Brulé properties to his oldest son M. S. Born at the family's White Hall plantation on November 3, 1814, at the age of twenty-four he followed the fairly common practice at the time in the South of marrying his first cousin Augustine Tureaud. The wedding took place at his bride's family's Union plantation in St. James Parish on January 25, 1838. Shortly thereafter, the young couple chose to reside at the Houmas plantation where M. S. planned to follow his father and grandfather's examples and build a grand plantation home.[111] After living for a while in a modest structure on the property, he began work on his great house. As the first phase of the project, he constructed two buildings separated from one another by a wide yard. They were designed to be the outbuildings of his manor house that was to be completed sometime in the future. One building contained a parlor, library, and dining room, while the second building contained the sleeping quarters. With the distance between the two structures approximately a half-block away, living at the estate during periods of inclement weather was not always comfortable or convenient. Unfortunately, the young Bringier's plan for a great house never came to fruition. With the coming of the Civil War, any chance of completing his dream house faded away forever along with the antebellum way of life and much of the Bringier family's fortune.[112]

Though he never completed his dream of building a grand home like his father and grandfather before him, M. S. did inherit his grandfather's love of entertaining. Despite the lack of a great residence to showplace, his Houmas plantation home became a center of hospitality and entertainment for guests and was frequently the site of large elaborate parties. When not entertaining or managing his plantations, M. S. dedicated much of his leisure time to scientific study and mechanical experimentation in the extraction of saccharine matter from sugarcane through the use of the diffusion process instead of the traditional method of cane-juice extraction by compression rollers. His studies and experiments were so successful that he obtained twelve different patents for his improved devices and became noted in U.S. government publications for his work on improving sugarcane production.[113]

His knowledge and understanding of the sugar industry was not limited to his scientific efforts. As manager of the Houmas and Brulé plantations, he demonstrated that he possessed the practical talent of mastering the sophisticated skills needed by a planter to operate a large sugar estate successfully in the antebellum period. Under his tenure, the Houmas and Brulé tracts became some of the most productive sugar acreage in Louisiana. Operated and equipped separately, both properties cultivated sugarcane and had their own sugarhouses. The only major difference between the two plantations was that Brulé did not have a residence house located on it. So skillful was M. S. Bringier's management and so great a financial asset did the estates become that an offer in 1858 of $575,000 from John Burnside to purchase the Houmas property was turned down by the family.[114]

Although Doradou's management-sharing plans established prior to his death helped to abate some of the issues that resulted from his dying intestate, more problems remained to be resolved. Understandably, the individual most affected by Doradou's death was his widow, Aglae. Despite being left with considerable financial resources, she now not only had to handle the emotional loss of her husband and children's father, she also had to deal with the complicated task of assuming the business and financial responsibilities associated with operating multiple large antebellum sugar estates. This was no simple task. Despite the fact that Doradou's death was a prolonged ordeal, as far as family members knew, he never discussed the impending outcome of his illness with anyone, even his wife, Aglae. Similarly, he made no legal preparation for events after his death. Perhaps it was his way of denying his mortality? Regardless of why he failed to take steps to protect his wife and children, Aglae was faced with a considerable challenge. Social norms of the period usually dictated a clear delineation of responsibilities among the members of a sugar-planter's family. Management of the estate's operations was largely the responsibility of the man of the family. Most women received little training either in school or by their husbands in the skills needed to manage a plantation successfully. Socially denied a political voice and clearly defined legal rights, many antebellum widows in the South found themselves at a great disadvantage while functioning in the role of a nineteenth-century sugar planter.[115]

Aglae had the good fortune to have had resources available to her that few other widows of the time could match. As discussed previously, her sister-in-law Fanny had taken over the management reigns from her husband Christophe of their Bocage plantation several decades prior to Doradou's death. Fanny's successful management of Bocage demonstrated that a female could master the varied skills needed to operate profitably a large sugar estate. Although Fanny Colomb died approximately two decades before Doradou, the example she set as the manager of Bocage estate provided Aglae with the valuable knowledge and the confidence to meet the challenge of running her husband's holdings. Additionally, Aglae benefitted from the assistance provided to her by her grown children and their spouses, including her talented son M. S. and her very successful sons-in-law Duncan F. Kenner, Martin Gordon, Jr., and Hore Browse Trist. Though she gave her eldest son M. S. power of attorney and most of the responsibility for managing the family's plantations, including White Hall, Houmas, Brulé, and the Hermitage, over the ten years that followed Doradou's passing there appears to have been considerable debate among the family members as to the best way of handling the many business issues associated with the Bringier holdings. Sons, daughters, and sons-in-law all seemed to have had advice—much of it conflicting—for Aglae on how she should handle her properties. The family was so divided over the estate's confusing legal situation that at one point Aglae's decision to sell some of her holdings to pay for debts was unsuccessfully challenged in court by two of her daughters, Mrs. Duncan F. Kenner and Mrs. H. Browse Trist, and their husbands. Although she toyed with the idea of liquidating her holdings to pay the family's debts, in the end Aglae followed the advice of those, including most notably her son M. S., who counseled that she keep the family's plantations in operation and work to reduce the financial claims against the family gradually. In spite of the legal issues and disputes, the family remained close. Until she died in New Orleans in 1878 at age eighty-one, Aglae, or as she was more often referred to in her elder years—"Bonne Maman" or "Bonny"—served as family matriarch who was loved and respected by relatives and friends alike.[116]

Fortunately for Aglae, with the considerable financial resources left to her by her husband and the assistance and advice of her family and friends, she successfully surmounted the business and social challenges brought on by Doradou's death. Aglae even managed to assume a role of prominence among the region's financial movers until she faced the ruinous effects of the Civil War. She was particularly generous to family members. According to family genealogist Trist Wood, Mrs. Bringier financed her daughter Myrthe and her husband Richard Taylor's troubled business activities for an estimated $300,000. Other family members benefited from her respected position in the business community. Such was the case when Martin Gordon, Jr., the husband of Aglae's third child Louise and a leading member of the business community in New Orleans, boasted in 1853 to a planter friend, who was attempting to put together a large business deal, that he could obtain the needed funding because of his ability to win Mrs. Bringier's support in the transaction which would "enable [him] to get all the money [he wanted] at [the favorable rate of] eight percent interest [from the area's bankers]." [117]

Chapter 4

The Hermitage

Among the common traits of colonial life shared by both the English colonies along the East Coast and France's colonial possessions along the Gulf Coast was that land ownership was the chief means to the accumulation of wealth. In Louisiana, agriculture was the principal economic pursuit of most residents. As was the case with Marius Pons' construction of his White Hall estate, when their agricultural pursuits proved successful and the colonial planters accumulated great wealth and fortune, they often used their fortunes to build large comfortable homes. The greater the wealth accrued, the larger and more elaborate the home.[118] To Marius Pons, his White Hall home was more than just a handsome residence for his family; it served as the showplace of his estate and became a visible grand symbol of his wealth and status in the colony. With the emergence of a successful sugar industry in Louisiana at the end of the eighteenth century, the region's planters accumulated great wealth, which they used to construct even grander manor homes. Though the style and size of the planters' homes varied from family to family and period to period, typically each structure was as impressive and as opulent as the owner could afford.[119]

The growing demand at the time for additional agricultural acreage resulted in a great deal of buying, selling, and transferring of property among the region's planters. With his estate in St. James Parish earning large profits, Marius Pons frequently participated in the real-estate market and acquired additional land holdings both in the area of White Hall as well as other investment properties for a possible resale or as sites where he could expand his agricultural ventures. Often Marius Pons and his colleagues would sell, trade, and reacquire huge tracks of land as if they were young boys trading small collectibles. One such real-estate transaction of Marius Pons involved the land that eventually became the home site of Doradou Bringier's Hermitage plantation.[120] On May 7, 1804, Marius Pons acquired from Pierre Part's estate a large section of property measuring twenty arpents wide and forty arpents deep that fronted on the east bank of the Mississippi River in Ascension Parish nearly opposite to the settlement at Donaldsonville. However, later on the same day he conveyed the property to J. C. Wederstrandt. After only five months, Marius Pons reacquired the land on October 10 for use by his fifteen-year-old son Doradou. In spite of his young age, Doradou developed a growing interest in the workings of a plantation. Upon learning of his father's sale of the Ascension Parish tract to Wederstrandt, he began to discuss with Marius Pons the possibility of acquiring his own plantation. Admiring his son's business sagacity, Marius Pons agreed to help Doradou secure his own property and provided him an advance on his inheritance of 1,600 *piasters*. Marius Pons then used the money to repurchase the estate previously sold to Wederstrandt. Holding the property for another two years, Marius Pons transferred the land tract measuring twenty arpents across the front and forty arpents deep to then seventeen-year-old Doradou on November 27, 1806, when the young man reached the age of majority. Sole ownership of the property did not pass to Doradou until Marius Pons relinquished his final share of ownership in 1812.[121]

When Doradou first obtained his share of the property, he made few improvements on the land. This changed when he married young Aglae and they decided to make the property their home. As his father decided earlier in regards to the design of White Hall, Doradou wanted his and Aglae's new home to be special and to befit his growing wealth and stature in the community. Hiring designers or architects to plan manor homes did not become common practice until a few decades after the Bringiers started to build their Hermitage manor house.[122] At the time work started on the house in August 1812, the use of professional designers in the region, though available, was rare. Most of the professionally designed stately mansions, which remain today in south Louisiana, were constructed in the 1840s and 1850s, several decades after the Hermitage. Despite the limited use of designers during the period and with existing records not identifying any particular individual, it is likely that Doradou and Aglae had some assistance in the planning and construction of their new home. It is suspected that his father Marius Pons had assistance in designing White Hall. Records also reveal that in his later years Doradou contracted the popular architectural firm of Dakin and Dakin for several projects. Because of the Bringiers' practice of using professional help to design structures before and after the construction of the Hermitage, it is likely that Doradou also had help in planning and designing his home.[123]

Other evidence suggesting that Doradou had professional assistance in the planning of the Hermitage includes the design of the structure. During most of the eighteenth century in Louisiana, the Creole plantation house was the predominant manor-house style. With the spread and success of the sugar industry in the area, its accompanying influx of wealth, and the arrival of large numbers of Americans following the purchase of the colony by the United States, new ideas about the design and construction of the region's stately structures began to surface in the territory. These ideas were brought into the area by both the newly arrived Americans, who carried with them knowledge of the classical designs common in the planter homes along the East Coast, and Louisiana residents, whose newly acquired sugar fortunes permitted frequent travel to Europe to bring home not only fine furniture and decorative arts but also new design ideas for their homes. The new design interest of the planters centered around the Classical-Revival style in architecture. When modified by local builders and craftsmen to meet the climate, materials, and ethnic demands of the region, the imported classical ideas soon evolved into the Louisiana-Classic style of architecture. With its dependence on its temple-like colonnaded porticoes, Greek-Revival styling soon dominated the plantation home design in much of Louisiana and other sections of the deep South. Although the Greek-Revival style grew in popularity, at the time Doradou and Aglae were planning the design of their new home, Louisiana's interest in the Classical-Revival style was at its earliest stage. It is likely that the young Bringier couple were among the very first in the region to include the classical style in the design of their home.[124]

Doradou spared little expense on the construction of his new home. Located on the east bank of the river across from Donaldsonville and approximately seventy-five miles upriver from New Orleans, work on the home took years to complete. At the time work on the structure began, the area suffered from difficult wartime conditions and shortages caused by the British blockade of the area's ports during the War of 1812. Because they encountered considerable difficulty in obtaining some of the materials needed to construct their home, the Bringiers were forced to rely on their own resources and skills for the goods and materials they needed. Artisans and workers on the estate produced nearly every item used in the building. With the exception of some native cherry wood, which was used in the construction of the interior stairway rail and spindles, lumber for the house was obtained largely from cypress trees felled on the plantation. According to family tradition, workers at the Hermitage work site produced the locks, hinges, woodwork, and nails. Also, according to family tradition, the only items supposedly not made on the estate were "special shells" imported from the West Indies for the making of cement and the glass used in the windows of the house. However, the current owners of the property, who have spent years in restoring nearly every aspect of the house, have found no evidence of any such "special shells."[125]

During the first months of construction of the new house, the newlywed couple lived at his family's White Hall estate. Although the couple got along well with the other family members, they desired to set up housekeeping in their own home as quickly as possible. To insure that the work on their future home met their specifications and design, Doradou spent much time traveling back and forth from White Hall to the new construction site. As work progressed, the couple decided to move to small temporary quarters at the Hermitage property. Though modest, their little house suited their needs while work continued on their future home. Living at the site enabled Doradou and Aglae to supervise carefully every aspect of the work on the house. The small structure was later moved to other parts of the plantation and used at different times as a guesthouse and the plantation's slave hospital.[126]

Although focused on the design and construction of his new home, Doradou did not neglect the rest of his estate. Realizing that additional land was needed for him to maximize the agricultural production of his plantation, Doradou worked to increase the size of his holdings. On August 26, 1812, just two months following his wedding and within days of the start of construction on the manor house, Doradou acquired the neighboring Amelung plantation, which greatly increased the size of his estate. Located adjacent to and upriver of the Hermitage property, the acquisition of Amelung nearly doubled the size of Doradou's estate along the Mississippi River to 33.75 arpents or an area of approximately forty acres. Over the coming years, he continued to acquire land, including a large tract called the great Houmas grant, which extended from the banks of the Mississippi River back to the shore of Lake Maurepas. He developed large sections of this tract into his Houmas and Brulé plantations.[127]

In spite of wartime shortages, Doradou and Aglae completed and moved into their new home shortly before Doradou reported for military duty in 1814. Among the earliest of the plantation homes in Louisiana to be built in the Greek-Revival style, the Hermitage presented an imposing classical appearance when first viewed by a visitor. However, upon entering, a visitor found an intimate interior in which the rooms were rather small in size when compared to those found in some of the grand plantation homes built after the Hermitage.[128] It was constructed so that large galleries encircled the family's living quarters. Supported by twenty-four massive Tuscan columns—a feature later copied by the builders of many of Louisiana's grandest antebellum houses—the gallery provided the house with a shaded surround and cool air during the hot summer months. The large heavy columns served as support for the second floor gallery and rested on short plinths to provide them level support. A decorative dentil course encircled the top of the house just below the massive hipped roof, which held two dormers that opened from the attic. The first floor of the structure was fabricated of solid brick walls, four bricks thick. The brick walls connected to a brick foundation, which flared out underground. Following a common building practice of the time in Louisiana, the first floor of the structure rested directly on the ground. On the bottom floor, four rooms opened to a center hall and stairway. The second level was constructed with a similar room layout as that of the first with a stairway and center hall connecting four rooms. Instead of solid brick, the walls of the second floor were made using the traditional building materials of the region of bricks between posts (*briquette entre poteaux*). Bays on both floors of the house were built with molded frames. The primary entrance into the house on the first level was equipped with two large wooden doors, whereas, the second level entrance was only a single transomed door. An additional single entry door was located on the right side of the structure's lower level. Shuttered French doors, designed to promote ventilation, protected windows on both levels of the house. Finally, chair rails and cornices protected the interior walls. The symmetrical placement of windows and doors as well as the floor plan of the house with its central hallway, inside stairs, the main reception rooms on the ground level, and the bedrooms on the second floor proved to be a popular and practical design and soon became a common characteristic of the many classically-designed-manor homes built after the Hermitage.[129]

    Climate conditions in the deep South greatly influenced the design of the region's manor homes. The semitropical climate of South Louisiana required that regional builders design their structures to provide as much protection from the weather as possible. Since the winters were usually short and mild and the summers long and hot, design concerns centered more on cooling a house than heating it. The thick walls and the shady and breezy gallery added to the Hermitage's comfort and were supplemented by two window dormers in the attic, which allowed heat to escape the large windows. Additional exterior openings extended down to the first floor in nearly every room. Some of the openings on the sides of the house were large windows while most others were larger openings that were protected by French doors and exterior shutters. The portals opened onto the shaded galleries and facilitated airflow throughout the house, with the larger openings serving interchangeably with the structure's doors as passageways and offered access in and out of the building. With only a moderate need for heating, the larger rooms in the building were equipped with fireplaces. Originally, the fireplace mantels were cypress in the common Adam's style of the Federalist period. However, sometime in the 1830s, Doradou and Aglae added to the house and remodeled their upstairs bedrooms. During the remodeling, they replaced the old mantels in the upstairs bedrooms with new imported black Italian marble mantles. On the ground floor, the Bringiers used a different design for the fireplaces. Because of their larger size, the fireboxes and their chimneys served as the focal point of the first-floor rooms. To accentuate their placement, Doradou and Aglae followed the prevailing regional practice of the time of surrounding the mantlepiece with a simple, yet elegant, decoration of painted wood adorned with minimal design and decoration. As with most structures of the time in South Louisiana, there was no basement in the house.[130]

Over time as the Bringier family increased in size, the house became crowded. To address the space shortage, Doradou sometime in the mid to late 1830s added to the rear of the house a two-story wing approximately sixty feet in length, with black marble mantlepieces and spacious galleries. The addition's centerpiece was a large room that over time was used as a ballroom, as the summer dining room with a slave-operated punkah, and as a storage area. The second floor was used as a billiard room and for additional bedroom space for Doradou and Aglae's growing family. Though the addition had a gallery held up by columns, the structure was not as grand or impressive as the original house. The upper galleries surrounded the addition and were of a shallower depth than those found on the original structure. The addition's gallery was supported on the lower level by simple brick pillars and square wooden columns on the second. Largely utilitarian in design, the addition lacked the grandeur of the original structure, yet it remained part of the house for nearly a century until it was torn down in 1932.[131]

In addition to the construction of the house's rear wing, over the years the family added several auxiliary structures to the estate. The operation of any large sugar plantation of the time necessitated the construction of support buildings. In an 1833 entry in his plantation journal, Doradou noted that in addition to the "master's house" the estate had grown to include a large sugarhouse (with two production lines with each including its own battery of kettles, rollers, and steam equipment), four large storehouses for maize and forage, two large structures forty-feet square (one of which served as a kitchen and the other being the Bringier's former temporary home which by then had been converted into the plantation hospital and an office), ten slave cabins, two pigeonniers, a dairy, and a smoke house. To protect the main house from heat and the possibility of fire, the Bringiers followed the common practice of the time of placing the kitchen in a separate but nearby structure.[132]

Upon completion of the house and with the end of the war with the British in 1815 and thus Doradou's military service, the Bringiers were able to concentrate on furnishing their new home. As with the design and presentation of their stately manor home, the interior furnishings of their house also served as an important symbol of an antebellum planter's status. Doradou and Aglae were determined to make the Hermitage a proud expression of their success and social standing. Unfortunately, the Bringiers kept few records of their furniture purchases and holdings. No doubt the home was furnished in the style of the day with large armoires, tables, chairs, and mirrors. Usually constructed of fine woods including mahogany, walnut, cherry, and the more common cypress, some of the fine furniture of the colony was imported from Europe and the East Coast, while other pieces were constructed in the region by talented local cabinetmakers. Additionally, from the time they moved into their new home, the Bringiers used fine artwork to add to the beauty of the house. Several of the paintings were gifts from Aglae's uncle, Bishop DuBourg. While on his many travels abroad, the prelate often purchased art and other objects that caught his fancy. He frequently shared his purchases with Aglae and Doradou, as well as other members of the DuBourg and Bringier families. The bishop obviously had a wide range of taste and a good eye for distinctive things—he even provided the Bringiers with toilet articles and rare laces that were appreciated and enjoyed by the ladies and girls in the family.[133]

Since the bishop was so involved in the decorating of the house and in so many other aspects of the lives of the Bringier family, it suggests the possibility, if not the probability, that the prelate worked with Doradou in developing the design of the Hermitage house. More than Doradou or other members of the Bringier family, the multi-talented Bishop DuBourg traveled the East Coast, particularly to Baltimore and Georgetown, in addition to Europe where he was exposed to classically-designed homes, as well as to the latest ideas in house design and the pattern books master builders often relied upon for the technical information used in their new projects. Regardless of the identity of the designer, the result of the determined efforts of Doradou and Aglae and the members of their workforce was a home of regal stature and beauty. So successful was the design of the Hermitage manor house that its classical design was soon copied by many of the builders of the region's grand manor-homes. Today, after nearly two-hundred years, the Bringier's stately home is listed in the National Register of Historic Places and is considered by many to be among the earliest Greek Revival plantation homes remaining in the state and has been called the most perfectly proportioned manor home located on Louisiana's historic River Road.[134]

Chapter 5

Life in the City

The Bringiers loved life at the Hermitage and their other plantations. However, in spite of the many amenities that plantation life held for the sugar planter and his family, life in the rural antebellum South presented limited opportunities for planter families to find alternatives to the routine of the plantation. Despite the fact that Louisiana's population had experienced steady growth from colonial times, by the dawn of the Civil War few urban areas were to be found in the state. If New Orleans and its surrounding faubourgs are excluded, in 1860 Louisiana possessed only a handful of urban centers, including Baton Rouge, Shreveport, Plaquemine, Homer, Alexandria, Thibodaux, Minden, and Donaldsonville. Of these, only Baton Rouge had a population greater than 5,000, with the others all having less than 2,500 inhabitants each.[135]

The closest urban community to the Bringier plantations was Donaldsonville. Located on the west bank of the Mississippi River where Bayou Lafourche intersected the mighty river, the town was only a short boat ride from the levee, which protected the Hermitage from the river. With a population of only 1,500 in 1860, Donaldsonville offered only minimal entertainment and cultural venues. On visiting the town in 1861, the noted British traveler William Russell described it as a place of "odd, little, retiring, modest houses."[136] The only structures of note at the time were the elegant Ascension Catholic Church, the parish courthouse, and the jail. In spite of its limited size, the town served as the region's commercial center and was the place where the Bringiers carried on much of the day-to-day business activities required to keep their plantations functioning in good form. Donaldsonville also provided the area's residents some simple diversion from their daily routines. Among the more popular activities offered were weekend horse races, frequent parades, reviews, balls offered by local militia units, several restaurants, and an assortment of merchandise outlets, including the shop "The Parisian Rougeau," where books were bound and popular novels and nonfiction works rented at the rate of $3.00 a year. The town also boasted of its own theater where both locally produced and traveling productions took place. The local newspaper, *Le Vigilant*, earned a reputation for its tireless campaign promoting a national literature in French. The efforts of the paper gave the region around Donaldsonville an extraordinary impetus in literary production by publishing a range of local, national, and international authors.[137]

Doradou and his family greatly enjoyed the rural life of their Hermitage estate in Ascension Parish. However, the Bringiers like many of their other wealthy planter colleagues in the region also spent considerable time in New Orleans. Regarded by many at the time as the "Paris of America," the myriad pleasures and luxuries of antebellum New Orleans offered the planter and his family a welcome respite from the tedium of life on a rural plantation. Business, schooling, shopping, and the enjoyment of the entertainment and social life of the big city were all enticements, which induced the Bringiers to spend extended lengths of time in the Crescent City.

While visiting the city, planters and their families either stayed at one of the city's splendid hotels or purchased homes. Doradou and Aglae preferred the comfort, consistency, and investment advantages of having a home of their own in the city. Over the years, they resided in several different houses in the most prestigious neighborhoods. Because Doradou's father, Marius Pons, had owned several properties in New Orleans, including a house on Bourbon Street between St. Ann and Dumaine, it is possible that Doradou and Aglae had use of these properties. However, it is known that among the family's earliest homes in the Crescent City was a house located on Esplanade Avenue. It is likely that the house was also owned and used by Doradou's father for his city home. The increase in the population of New Orleans during the colonial period resulted in many of the wealthy Creole families abandoning the crowded Vieux Carré for the trendier neighborhood along the boulevard, which boarded the down-river side of the old quarter of the city. Esplanade Avenue, named for the parade ground of the old Fort St. Charles which once stood near where the street met the river, was appealingly landscaped with sycamores and oaks and was for the time the location of some of the grandest homes in New Orleans. For years, it was the fashionable place for the wealthy Creole families of the city to reside.[138]

Members of the Bringier family, including Marius Pons' son Don Louis and his daughter Louise, lived on Esplanade Avenue for decades. Even though he continued to own property on the boulevard for years, Doradou's ever-wandering business eye led him to the upriver side of the Vieux Carré for a new and grander city home for his family. Suburban

Life in the City

growth along the Esplanade encouraged developers to do likewise on the upriver side of the city. Originally part of the commons area which separated the old city from the American populated suburb or Faubourg St. Mary, the area upriver from the Vieux Carré had been kept open for a public avenue and a canal. Eventually the area became known as Canal Street, though no waterway was ever built on the land. The unusually wide median, which had been originally reserved for the location of the canal, made the street one the most attractive residential and business areas in New Orleans and the South.[139] In 1836, Doradou acquired for $8,850 part of a large building on Canal Street between Dauphine and Burgundy streets known as "The Three Sisters." The structure was so large that one section was eventually converted from residential use to become the Varieties Theater, and later the Grand Opera House. Though an imposing size, the structure was not comfortable and did not adequately meet the needs of the Bringiers. Their stay in the building was short. After only four years of residing on the popular Canal Street, Doradou looked elsewhere in the rapidly growing community for a new more comfortable home for his family.[140]

Whether it was because of the increased commercialization of the Canal Street area or just a case of the Bringiers finding a house they preferred to their cavernous Three Sisters building, Doradou and Aglae moved their family to an imposing structure located a little over a mile upriver from Canal Street known as Melpomene. Located on Apollo Street, which eventually became Carondelet Street, Melpomene became loved by the family and remained the Bringier family's city home for decades.[141]

The choice of the location of their new home upriver from the Vieux Carré tells much of Doradou's character, style, and business sense. Melpomene was located near the center of the city's American sector. Since the time Americans began arriving in New Orleans in large numbers there existed a high degree of tension between them and the Creoles who looked upon the newcomers as interlopers. The political, social, religious, business, and cultural competition between the two ethnic groups became so intense that by 1836 the state general assembly took the unusual step of dividing the city into three largely self-governing municipalities—the First Municipality being the Vieux Carré and predominantly Creole, the Second Municipality constituted the predominantly American sector upriver from Canal Street, and the Third Municipality being the remainder of the city below the Esplanade which was mostly settled by Creoles and non-Anglo immigrants. Hence, when

Doradou and Aglae purchased their new home on Carondelet Street, they ignored the social order of the time and the fact that the majority of their Creole friends and colleagues in the city had largely isolated themselves from having contact with the city's ever-growing number of Americans. Doradou and Aglae's move to the American sector was demonstrative of their somewhat unique open-arms acceptance of their American neighbors. Over the years, their friendly relations with Americans would be a major influence and factor upon the social and business lives of the Bringier family.[142]

The actual construction date of Melpomene is not known. Originally built sometime in the early 1820s by Joseph Theodore Bauduc, final work on the structure was completed after Doradou acquired a one-sixth interest in the property from his wife's brother-in-law, Seaman Field in 1838.[143] Three years later when Field faced financial difficulties, he sold Doradou a one-third share of the estate and the next year in 1842 the entire property.[144] Though purchased by Doradou for his family's city residence, at the time Melpomene actually lay in an area that was still largely a pastoral setting. Originally, the large plot of land where the house sat was connected to the river by an avenue of trees. However, with the American sector expanding rapidly on the upriver side of the city and with the American community beginning to rival the French sector in wealth and political influence, it was obvious to Doradou that the area around Melpomene would eventually develop into prime real estate. In spite of the area's growth and development, which included the construction of many new dwellings and farms, the region continued to maintain its rustic character for years. With few walkways and developed streets, family members were forced to rely on the family carriage and driver to take them "to town" to shop and visit friends on Canal Street and in the other sections of the city. As late as 1849 the area continued to be populated with sizeable pastures and empty lots of scrub trees and bushes.[145]

Constructed at a time when the area was still largely rural in nature, the house was built in the popular French colonial plantation style with double long two-story rear wings attached to it. Its hipped roof was very steep, with its eaves extending over the house's wide galleries on three sides of the building. The house's shuttered windows were the large casement type with multiple lights, which allowed for cooling breezes to pass through the house's upper level. At the time the Bringiers acquired the property, the main house was raised off of the ground by great square pillars. With the entire region subject to flooding from an overflow from the area's many waterways and poor drainage system, the construction of homes raised high above the ground was a common and practical architectural design throughout South Louisiana, which dated to the time of the earliest French settlers in the region. Once the construction of levees and drainage ditches decreased the danger of flooding, many of the owners of the raised-above-ground structures converted the areas beneath their houses to closed-in bottom-floor living spaces. The Bringiers made this modification to Melpomene by paving the downstairs area with great flagstones. The family used the area as a popular cool retreat from the heat and humidity, which afflicts New Orleans during much of the summer. As the family's space needs increased, the lower portion of the house was eventually enclosed by the Bringiers and converted into extra living space, which included new dining and drawing rooms.[146]

Though the newer levees improved the flooding situation around Melpomene, as even modern residents of the city today can testify, they by no means eliminated the problem. In May 1849, heavy rains and a crevasse caused such flooding around the house that family members were forced to use boats to travel to the city proper. Fortunately, the house was spared damage because of a series of seven banquettes (raised earthen walkways) that had previously been constructed and separated the house from the floodwaters which flowed a short distance from the house.[147]

Even for its time when large plantation homes were common, the imposing Melpomene was considered remarkably large for a family residence. Toward the end of its existence near the turn of the century, an architect estimated that the lumber contained in the roof and rafters of the house alone was sufficient to construct four large-frame dwellings. The layout of the house followed a typical design for homes of the period in which it was constructed. A wide hall lighted by a splendid skylight constructed of many smaller stained glass windows connected the many large comfortable rooms of the mansion. Like the Hermitage, the house was decorated with artwork, including deluxe books, silver and bronze sculptures, paintings, and engravings. Many of the art pieces the family used to decorate Melpomene were provided to the Bringiers from abroad by Bishop DuBourg during his extended stay in Europe after he gave up being bishop in New Orleans in 1826 and before his death in 1833.[148]

The elegant villa occupied the entire square bordered by Melpomene, Carondelet, Terpsichore, and Baronne streets. Although much of the property near the rear of the house was paved with large stones, the remaining area of the estate was planted with many types of trees and plants, including oranges, magnolias, and oaks. By the time of its destruction near the end of the century, the house and property were completely shaded by several enormous oak trees, which had originally been planted decades earlier by Martin Gordon, Sr., the father of Doradou's son-in-law and one of his business partners who had at one time owned a share of the estate with the Bringiers.[149]

An interesting family story maintains that one of the earliest fire engines brought to New Orleans was for some time housed at Melpomene in the vacant space under the raised structure. It was customary at the time for the young men of the city to organize into voluntary fire brigades. Doradou's son-in-law, Martin Gordon, Jr. (nicknamed "Bob"), was prominent in the local fire company and had successfully worked with his father in Philadelphia to obtain the new engine for New Orleans. Needing a convenient place to keep the machine at the ready, Gordon convinced the Bringiers to allow the stationing of the engine under their house.[150]

Martin Gordon, Jr., obviously was attracted to the stately house for more than just a convenient location to store his fire engine. His father, a friend of the Bringiers, at one time owned a one-sixth share of the house at the time that Doradou first bought an interest in the property from Seaman Field. The relationship between the Bringiers and Gordons evolved to a point where Doradou gave the hand of his fifteen-year-old daughter Louise Francoise to Martin Gordon, Jr. Married in 1835, the couple resided in New Orleans first at a house on Esplanade Avenue provided to them by her father and later on Royal Street in the Vieux Carré. Never losing his love for the majestic Melpomene house, in 1847, when he and Louise faced financial ruin because of a series of failed speculative business ventures he had made with his father, they readily accepted an invitation by Doradou to move into the house. Unfortunately, the event was not a happy one for at the time of the offer, Doradou was nearing the end of his battle with cancer and obviously wanted to have someone he trusted move into the home to help his wife Aglae manage after his passing.[151]

Shortly after offering Melpomene as a residence to the Gordons, Doradou died on March 13, 1847, while seeking medical care in Memphis, Tennessee. His death caused considerable disorientation for a time among family members as to the future of Doradou's properties and business ventures. Aglae was so distraught over Doradou's death that for sometime, she avoided both her city and country homes. Instead she made extended visits to friends and relatives in the Baltimore area and at her children's estates—Nanine and Duncan Kenner at their Ashland plantation, Myrthe and Richard Taylor at their Fashion estate, and Octavie and Allen Thomas at their New Dalton home. Although she had previously spent the majority of her time before her husband's death at the Hermitage, once she settled into her new life of widowhood, Aglae preferred life in the city with its conveniences to that of the plantation and as such made Melpomene her principal residence. For several years, the large house thus became the home of two distinct households—Aglae and the Gordons. Eventually, the Gordons regained their financial footing and decided that they wanted a new home of their own. However, instead of looking elsewhere for their new homestead, Aglae allowed them to use a section of her Melpomene property, which faced Baronne Street (Bacchus at the time) near Terpsichore to build their new residence. For many years thereafter, the large square of land was populated only with the Gordon house and its small outbuildings and the Bringiers' stately Melpomene with its wings, stables, and gardens dotted with its collection of hardwood and fruit trees.[152]

The two homes on the Melpomene property earned reputations for hosting some of the city's most elaborate dinners and social galas to take place in antebellum New Orleans. In addition to the notable families of New Orleans who were guests at the Bringier and Gordon homes, over the years the elegant Melpomene was visited by some of the South's and the nation's most celebrated personalities of the time, including Pres. Andrew Jackson, Confederate Pres. Jefferson Davis, generals John Bell Hood, Edward Canby, Braxton Bragg, Dabney Maury, Gov. Thomas Overton Moore, U.S. Secretary of State Hamilton Fish, Bishop DuBourg, legal crusader Myra Clark Gaines and Mexican-War diplomat Nicholas Philip Trist. Even after the Bringier children were grown and moved out of their parent's home, Mrs. Bringier and the large city house were seldom without visitors. With such a large family as the Bringiers, dinners at the home with as many as fifty family members were common.[153]

The Gordons also enjoyed entertaining. After he regained his fortune and built a new house on the square with Melpomene, Martin and Louise Gordon's home soon acquired a reputation as a center of hospitality in the city. Family members long remembered one particular Gordon fete that took place in the winter of 1856. At the ball-supper, the table was elaborately set with flowers and foliage decorations. Amidst the flowered decorations were dishes on which game were set as if they were in their natural setting. As the guests gathered around the table, Gordon signaled the waiters who suddenly lifted the feathered skins from around the game revealing sumptuously cooked birds, ready to be served to the amazed guests. To carry out this special bill of fare, Gordon went so far as to engage a special taxidermist in New York to make preparations for the affair. Martin Gordon, Jr.'s parties were so renowned that even a decade after his death, the *New Orleans Times-Democrat* quoted his granddaughter's boast that "Dinners were given by this prince of entertainers which, for their lavish and prodigal splendor, have become famous in a land renowned for the extravagance and generosity of its people."[154]

The large house also was frequently the site of family celebrations and special events. For example, in February 1851, Melpomene was the site of the wedding ceremony between Doradou and Aglae's seventeen-year-old daughter, Myrthe (Mimi), to Richard Taylor, the future Confederate general and war hero and the son of the late Pres. Zachary Taylor. Additionally, the house was the preferred birthing site for many family members. It was the custom for pregnant "daughters of the house"—even those who resided on estates some distance from New Orleans—to move to the spacious city house as the time of a birth drew near in order for them to receive the better medical attention afforded in the city. Due to the popularity of the practice, Gen. Richard Taylor who was visiting with his wife complained somewhat loudly that the stately house was being made into a "lying-in hospital." When an unidentified sister-in-law—whose being with child was obviously the immediate object of his remarks—overheard him, she informed the general that he was obviously laboring under the delusion that the house was his own and not her mother's. Needless to say, the military hero was dutifully humbled and immediately made the appropriate courtesies to apologize for his unfortunate sarcasm.[155]

The old mansion remained a center of the Bringier family's activities in New Orleans until the death, from what her doctor called "general debility," of eighty-one-year-old Aglae Bringier at the house on June 11, 1878. For a while after Aglae's passing, the structure continued to be occupied by Col. Robert Wood, a grandson of Zachary Taylor, and his wife Wilhelminia Trist and their children.[156] When they eventually moved from the old house, it remained vacant for sometime. The grounds were neglected and the entire lot and structure became overgrown with plants and weeds. Efforts to rent the building were largely unsuccessful for few individuals wanted to tackle the task of maintaining such a large old house with its grounds in such disrepair. At one time, the old house was leased for a short period to a Judge Howe and on another occasion it was used as a school for African-American children. There was even some talk that the city would take over the mansion, establish a museum, and show the place as a typical example of an old plantation home. Unfortunately, this, as did all other efforts to save the historic structure, proved fruitless. Gen. Joseph Lancaster Brent, the son-in-law of Nanine and Duncan Kenner, sold the grand old mansion, which had hosted over its life some of the city's most elaborate social events and had welcomed so many luminaries.[157] Put up for sale in 1891 with the approval of Aglae's daughter and the family's surviving granddame Mrs. Nanine Kenner, the old majestic house, whose roof was said to tower like a hill over the surrounding level of modern houses, its outbuildings, along with the many trees which shaded the property were razed. In their place, the property was subdivided into smaller lots where new smaller houses were built and sold.[158]

Chapter 6

Land and Livelihood

With their stately Melpomene city house and their multiple country plantations nestled along the Mississippi River, the Bringiers enjoyed a standard of living and wealth that the vast majority of both white and black inhabitants of Louisiana and the South could only dream of experiencing. Yet their less-fortunate neighbors in rural Louisiana, their sugar-planter colleagues, and the Bringiers did share at least one important life factor together. The climate, weather, and geography of the region and the vagaries of the cultivation of their crops determined much of the routine of their daily lives and activities. In spite of their many successful investments in real estate and other business ventures, sugar was the foundation of the Bringiers' wealth from which all else derived. The production of a successful sugar crop on a large estate was a complicated, expensive, and difficult task. All activities of those who lived and worked on a sugar estate revolved around the crop's needs. Because of the many variables in the growing and making of sugar, there was little idle time on the well-managed sugar plantation. When the crop itself did not demand attention, there were many other secondary activities such as the gathering of firewood, the cultivation of secondary crops, and the cleaning of drainage ditches that kept all hands on the plantation busy throughout the entire year.[159]

The nature of sugar cultivation often resulted in overlapping crop seasons on the Bringier plantations. Although on larger estates cane was usually planted more than once a year in different fields, a plantation's primary new crop was typically planted shortly after the end of the grinding season. The all-important grinding season marked the point when the cane was harvested and the actual production of sugar began. This usually occurred in late December to early January. Once the harvesting of the previous crop was completed, the fields were thoroughly plowed to make ready for the planting of a new crop. Because of the nature of cane, planters did not have to seed all of their fields every year. In Louisiana, a second and sometimes a third year of production could be obtained from the rejuvenated growth from the stubble left in the field following the harvesting of the previous crop. After a third harvest, the roots were plowed up and the field left fallow for nutrient replenishment. Planters usually supplemented this process by planting cover crops such as legumes and by leaving the rubbish and trash from the earlier harvest on the ground. When it came time to use the field again, the cover crops were plowed under, the rubbish was piled up and burned, and the ashes were spread across the ground to help fertilize the field.[160]

The planting of the new cane was a laborious undertaking, which absorbed a larger portion of a sugar plantation's workforce's time and effort than any other activity. The soil of the fallow fields was cultivated with rudimentary plows controlled by field hands and pulled by oxen, horses, or mules. Once large rows were laid out in the plowed fields, a small furrow was cut in the row's center and a piece of seed cane was dropped in and covered with soil. To accomplish this task, large carts packed with pieces of seed cane and usually pulled by oxen moved across the prepared field while workers took cane from the cart and placed it end to end in the furrow and covered it with dirt. Depending on weather conditions, the task was usually completed by early March. On occasion, monsoon-like rains and uncharacteristic cold spells could delay the process and force the field hands to work under miserable conditions. One such incident occurred in 1852 on the Bringier plantations when a cold wave struck the area with such severity that the planting process was forced to stop for days. On January 13, an unprecedented snowfall of four inches blanketed the Ascension Parish area and did not melt for four days. To compound the situation, within a week of the snow a second and even more severe cold wave hit the area. It was so cold that W. G. Wade, the overseer on Duncan and Nanine Kenner's Ashland Plantation a few miles upriver from the Hermitage, bemoaned in his journal that the wintry blast was the "coldest day that was saw in Louisiana."[161]

With the planting of the new seed cane completed and the arrival of spring, the focus of work on the Bringier plantations became the cultivation of the crop. Because young cane could be threatened and choked by the growth of weeds, groups of field workers or "hoe gangs" made up of both male and female workers carefully hoed away the weeds along the rows of new cane. By late June or early July, the crop usually reached a level when it was no longer jeopardized by weeds and was said to be "laid by." The weed cutting could now be eliminated in the cane fields. With the needs of the cane crop temporarily addressed, work shifted to the other tasks needed to keep the estate in shape for the fall's harvest and grinding season, including the repairing of the buildings and equipment, and the never-ending plantation chores of collecting firewood and levee repair. Perhaps the most important task addressed during the "lay-by" period was the maintenance of the estate's drainage system. Adequate drainage was a necessity for a successful sugar crop. Though the cane thrived in the damp fertile alluvial soil of the region, it did not survive long in wet soil. The mostly flat geography of south Louisiana's river region made for a sluggish natural drainage system. To address the problem, the Bringiers and other sugar growers implemented grids of drainage ditches and canals across their cane fields which drained the water from the fields and deposited it in the swampland located at the back of the many sugar plantations located along the Mississippi River.[162]

During the "lay-by" and other periods of the year, planters frequently assigned their workforce to the cultivation of crops other than the plantation's primary money crop. The Bringiers relied on secondary cash crops to supplement the income from their main crop. Although the most common of the supplemental agricultural items produced on the area's plantations included such items as butter, hay, orchard products, honey, oats, beeswax, market-garden vegetables, rye, wine, wheat, hops, barley, clover, grass seed, cheese, millet, and rice, the Bringiers developed a profitable supplemental crop of tobacco. Planters also depended on supplemental food crops to help defray their operating costs by producing products they could both sell to others and use in the operation of their own estates. Most notable of these products were those which served as food for the master and his family as well as the plantation's workforce. Commonly grown food crops included corn, peas, beans, sweet potatoes, and Irish potatoes. Additional agricultural products were produced for use by the plantation's eclectic collection of livestock, which included horses, milch cows, oxen, swine, sheep, and mules.[163]

October typically marked the beginning of the sugar harvest. Called the "rolling season," it marked the busiest time of the year for everyone on the plantation. From the beginning of the harvest until the sugar-making process was completed in either late December or early January, there were no inactive hands and very little free time for anyone on the estate. Despite the hard work that the harvest brought to the inhabitants of the plantation, everyone eagerly welcomed the harvest and the startup of the grinding and manufacturing process. The harvest commenced with the plantation's workforce being divided into gangs with specialized responsibilities. Teams of cane cutters consisting of both male and female workers were sent into the fields where they performed the backbreaking, repetitive operation of separating the cane stalks from their roots, stripping the leaves from the cane, and cutting away the useless unripened joints. Other gangs, sometimes made up of the elderly and children of the plantation, followed the cutters, collected the cane stalks, and transported them in carts to the mill.[164]

It was essential that the cane harvesting be completed as soon as possible. Not finishing the process by mid-December increased the possibility of the crop being damaged by the arrival of a severe cold front. This was the case in late December 1855 when Amedee Bringier reported in frustration to his wife Stella that at the Hermitage, "It has been utterly impossible to press canes through the mill since Monday. God only knows when I'll get through this cursed grinding. The whole field is nothing but ice. The pond and even the water in the well has . . . iced."[165] Despite the danger of an early cold snap, the speed of the cane cutting was dependent on the speed of the grinding and manufacturing process. The estate's mill generally set the pace of the cutters. Once a stalk of cane was cut, it needed to be processed fairly promptly because the juice in it would begin to spoil within two days. Hence, once the mill was started coordination between it and the field operation was essential. Planters had to insure that there was a sufficient supply of cane on hand for the mill to process and work efficiently, but not enough that would lead to spoilage of unprocessed cane. Because of the difficulty of shipping the bulky cane any distance, on most large plantations the mill was located on the property so that it could be reached easily from the estate's cane fields. Each of the Bringiers' plantations maintained separate sugarhouses and mills. At the Hermitage, the facility was positioned approximately 800 feet upriver from the family's big house, placing the large structure 200 feet closer to the river than the family's home. This placement made it easier for the Bringiers to move the large hogsheads of finished product to the river for transport. The size and design of the sugarhouses changed over time to meet the production needs of each of the family's plantations. Although specifics on the size of the sugarhouse are lacking, family records do reveal that in 1840 the sugarhouse at the family's Brulé plantation was 150 feet wide by 315 feet long or 47,250 square feet in size. Whereas, at approximately the same time, Duncan Kenner on his large Ashland estate maintained a sugarhouse of only 140 by 45 feet or 6,300 square feet.[166]

Over the years, the design of sugar mills changed dramatically, with many of the most significant improvements coming during the final decades of the antebellum period. As was the case at White Hall under Marius Pons and the early years at Doradou's Houmas and Hermitage plantations, early sugarhouses were basic in their design and function. Little more than sheds, the early mills consisted of devices for crushing the cane and extracting and boiling the sweet juice until it became a thickened sugar product. Gradually over the years, the process and machinery became far more sophisticated and expensive. In his journal of 1833, Doradou noted that his Hermitage plantation stretched thirty-three and three-quarter arpents along the river and included a sugarhouse with steam engines for processing the sweet crop. He also noted in his journal that the plantation had a workforce of sixty workers, twenty children, and an array of farm animals, including twenty horses, forty oxen, and fifty cows.[167]

As was the case at the Bringier properties, it was typical on large sugar plantations that the mill was located in the sugarhouse. The outfitting of this facility was among a sugar planter's most expensive investments. Over the years that sugar was produced on the Bringier properties, the cost of the equipment used in the production of the crop increased dramatically. When sugar was first introduced on the family's plantations, approximately $10,000 was needed to outfit a mill. As the industry matured and improved production techniques and machinery were introduced, the average price for equipping a sugarhouse jumped to $50,000, with some of the larger and more sophisticated units costing as much as $100,000. In spite of the cost, the Bringiers believed in the use of technology as it developed and consistently updated their equipment so that their production facilities largely remained state of the art. However, the state of technology development in the sugar industry during the first half of the nineteenth century was such that there was a great deal of hit and miss efforts in improving sugar production. Family members and colleagues who were involved in the industry frequently collaborated in their efforts to identify the most effective new and improved equipment and techniques in sugar production. This was the case in 1846 when Doradou and his son-in-law Duncan Kenner worked together in purchasing new sugar-making equipment for the Hermitage sugarhouse. Doradou spent $7,500, a price Kenner admitted was more than other available equipment, because of his son-in-law's observation that the "mills and engines [were] the best" because "they [were] more substantial [with] more iron out and in and [are] better finished."[168]

With their Houmas plantation having $75,000 and the Hermitage $25,000 worth of farming implements and sugar-making machinery, by

1860 the equipment on the Bringiers' properties in Ascension Parish ranked second in value among sugar growers in the parish. Although the purchase and maintenance of an estate's sugar-making equipment were a grower's single largest expenditure, the operation of a large sugar plantation required an investment in many other costly items. As noted elsewhere in this work, the cost of acquiring and maintaining a plantation's slave force was also among a planter's greatest expenses. Other major costs involved the production of food and other subsistence products for use on the plantation as well as the acquisition and upkeep of livestock and additional acreage for future expansion and use. A list of tools noted in an inventory of implements used on Duncan Kenner's Ashland plantation was typical of those used at the Bringiers' other plantations. Among the tools included on the list were such items as plows, carts, cultivators, spades, shovels, sickles, mowing equipment, tarpaulins, hayforks, axes, scythes, and dirt scrapers. Additionally, plantations maintained a variety of livestock for use as both a source of food and as work animals. By 1860, the livestock alone at the Hermitage was valued at $10,500 and included ten horses, sixty mules, five milch cows, four oxen, twelve cows, and fifteen sheep. At the larger Houmas/Brulé plantation the size of the livestock herd was similar to that of the Hermitage, with ten horses, ten milch cows, eight oxen, ten sheep, and ten cows. However, the 170 mules employed at the larger Houmas/Brulé property were nearly triple the number used at the Hermitage and was approximately equal to the 173 mules used by Duncan Kenner at his Ashland plantation.[169]

Once the cut cane was delivered to the sugarhouse, the grinding or milling process began in earnest. It was a time of great activity and excitement on the plantation that typically continued nonstop until all of the estate's cane was processed, a period usually of ten to twelve weeks. Although the addition of new technology over the years enhanced the process, the transforming of sugarcane into raw sugar has remained basically the same since sugar was first introduced into Louisiana more than two centuries ago.[170] On the antebellum sugar plantation after the newly-cut cane was crushed by a series of large rollers, the juice was collected in vats located below the rollers. Next, the juice was filtered to remove pulp and any other foreign matter, which may have fallen into the vats. The juice then was processed through another series of kettles. Typically a plantation used a set of four open-air kettles called the "grande," the "flambeau," the "syrup," and the "battery." The kettles were laid out from the largest to the smallest. The juice was then boiled and the water in it evaporated. As it thickened, it was transferred to the next vat in line. The process continued as the thickening juice was transferred from vat to vat until the boiling juice became syrup in the "battery" kettle at the end of the line.[171]

After the juice was reduced to syrup, the "strike," the most important step in the sugar-making process, occurred. An expert known as the sugar maker was needed for this step. Usually an itinerant worker hired by a planter to bring in a crop, he determined when the boiling syrup reached the precise temperature and consistency to attempt a "strike"—the point when sugar crystals began to be produced indicating that granulation was near. It was at this moment that the sugar maker ladled the hot syrup into the cooling vat where it crystallized into sugar. If the "strike" did not take place at the right moment, the syrup would cool and become molasses, which rendered a crop almost worthless. Once cooled, the raw sugar was placed into large wooden casks known as hogsheads. When first packed a hogshead weighed approximately 1,100 pounds. After the liquid molasses was drained from the container and placed in smaller barrels with capacities of forty to fifty pounds for either use on the estate or for market, the weight of the hogsheads of sugar shrank to about 800 pounds.[172] On the Bringier plantations, the filled hogsheads were stored in the various sugar warehouses located on each of the properties until ready for sale. At that time the product was typically moved to each estate's dock. Houmas and Hermitage, like most of the large sugar estates of the time, were located on the Mississippi River or some other navigable waterway, which made shipping their products to market a relatively easy operation. From the dock the filled hogsheads were loaded on schooners, including the vessel *Augustin*, or more frequently on one of the many shallow-draft steamboats, such as the steamer *Creole*, which plied the Mississippi and other waterways of Louisiana transporting people and materials to and from plantations and to the state's urban communities.[173]

When the crop reached the city, it was placed in the hands of the planter's factor—an agent who was authorized to sell goods or purchase items in the planter's name. The marketing and selling of a crop were nearly as important to the planter as was producing the crop itself. If not marketed properly, shipping and storage costs could deflate a planter's profit on his season's crop. Compensated by charging a percentage of the totals of the goods they bought and sold for planters, factors frequently faced criticism from Southern planters for being outrageous exploiters. In spite of the sometimes rocky relationship between planter and factor, the planter's agent played a vital and specialized role in the economy of the plantation. Their interdependence and reliance on one another often led to special relationships developing between planters and their factors. The Bringiers' relationship with their factors was indeed special. During their early years at Hermitage, Doradou and Aglae used the services of their brother-in-law and one-time partner in their Melpomene estate, Seaman Field. In later years, they continued to keep their factoring business within the family when they used the services of their daughter Louise Francoise's husband, Martin Gordon, Jr. One of the region's most successful factors, he preferred life in the city to that of the plantation. Thus, he and his wife made their residence at the Bringier's Melpomene property in the Crescent City. In addition to serving as the commission agent for Doradou and Aglae, Gordon also provided factoring services to several other members of the family including brothers M. S. and Amedee Bringier and his other brothers-in-law Duncan Kenner, Richard Taylor, Allen Thomas, and Hore Browse Trist.[174]

The quest by planters to increase the size of their crop production was never ending. Even the disastrous consequences and difficulties brought by the Civil War to the industry did not stall planters' efforts to improve the quantity and quality of their crop. Bringier family members actively worked to develop new production techniques and equipment for the sugar industry. Duncan Kenner led the family efforts to bring reform to their industry. He was the first sugar planter to use a portable railroad as he did on his Ashland and Bowden plantations to transport efficiently cane stock to his mills. He was also among the first in the industry to adopt the hydraulic pressure regulator. These advances were so effective that both devices became commonplace on sugar plantations by the end of the century. However, Kenner's most important contribution to the sugar industry was his effort to awaken Louisiana's planters to the need and benefit of working together to enhance their industry through the formation of professional organizations and agencies. His effort culminated in 1877 with the formation of the Louisiana Sugar Planters Association. Created with the specific mission of promoting the culture of cane and the manufacture of sugar, the organization rewarded Kenner for his tireless efforts to improve the sugar industry by electing him the first president of the group, a position he held continuously until his death in 1887.[175]

In addition to Kenner's efforts to improve the sugar industry, Bringier brothers M. S. and Amedee also worked for years to improve the production of sugar cane on the family's plantations. Focusing on ways to improve the implements used on the plantation to grow and process the cane, M. S. and Amedee spent years working on a series of inventions. Their endeavors were so successful that over the years the two brothers received a combined total of fifteen patents for items designed to improve the sugar industry. Of the two men, the elder of the two brothers, M. S., received twelve different patents (see Appendix B). Obtaining his first in 1859 for an improved steam boiler, M. S. continued to patent new or improved equipment for the industry until he obtained his final patent in 1876 for an apparatus for extracting saccharine liquor from cane. His younger brother Amedee did not obtain his first patent, which he jointly acquired with N. B. Trist, until ten years after his brother obtained his initial one. Eventually obtaining a total of four patents, Amedee continued to work on inventing new equipment and tools until 1884 when he obtained his fourth and final patent for a device called the Bringier Pulverizing Cultivator, which he marketed with the help of Duncan Kenner.[176]

Throughout the antebellum period and taken as an aggregate, the Bringiers remained among the largest and most successful sugar producers in the nation.[177] Between 1829 and 1845, the sugar production from their Hermitage and Houmas/Brulé properties yielded handsome average gross proceeds of $57,000. However, the disruption caused by Doradou's death two years later in 1847 greatly affected the family's financial status. Although the Bringiers' properties continued to produce successful crops which yielded even greater income, the disruption in the plantations' management caused by Doradou's passing resulted in an indebtedness for his wife Aglae of approximately $180,000 and an even more ominous problem of a reduction in the availability of credit—a necessity for the successful operation of any large agricultural endeavor. The legal status of Doradou's estate remained unsettled for several years after his death. However, wise management on Aglae's part with the assistance of the members of her family, including her sons and sons-in-law, along with several years of bumper sugar crops on the family's three crop-producing units at her Hermitage, Houmas/Brulé, and White Hall properties successfully revitalized Mrs. Bringier's financial fortunes. Her 1852-1853 sugar crop alone totaled more than 2,600 hogsheads and was valued between $116,000 and $145,000—making Aglae the largest sugar producer in the state for that year. Indeed, in the three-year period between 1853 and 1856, the Bringiers' Houmas property led all other plantations in Louisiana in sugar production.[178]

Nevertheless, in spite of her financial recovery, her personal situation became so crucial that she seriously considered selling all of the family's plantations. In addition to the large debt noted above, the inheritance and legal problems associated with the fact that her husband left no will along with the difficult chore of operating several of the region's largest plantations weighed heavily on Aglae. In 1858, over a decade after her husband's passing and as her frustrations mounted, Aglae decided to sell all of the family's plantations as a means of settling Doradou's debts and succession issues. She hoped the sale would end the uncertainty and debate among family members concerning Doradou's succession.[179]

With the support of some family members, she carefully obtained appraisals of each property and made the determination that she would not sell any property for less than the appraised value. She also decided, as it was the most valuable of the holdings, that the Houmas plantation should be sold first. Although her son-in-law Duncan Kenner and Stella Bringier, the wife of her son Amedee, looked into making bids on the Houmas and Hermitage properties respectively, the decision was made to put the properties up for public auction. However, the only serious interest in the properties came from the Bringier's new neighbor John Burnside, who had recently purchased the Houmas House plantation and was looking to add to his holdings. Burnside offered $575,000 for the entire Houmas tract, a price $25,000 less than what Mrs. Bringier demanded. Aglae rejected Burnside's offer and bought the properties herself when no other bids were received from the general public. Taking this unusual step resolved the inheritance issues associated with Doradou's death and gave Aglae sole legal ownership of her late husband's properties. The value of her purchase was approximately $776,000 and was backed by Duncan Kenner who endorsed her promissory notes on the deal.[180] Once she possessed full ownership of the properties, she began to take steps to share possession with her older sons, with M. S. obtaining a share of Houmas/Brulé and Amedee a share of the Hermitage. The booming sugar economy of the time quickly converted the Bringier brother's ownership shares into highly profitable ventures. By 1860, a couple of years after Aglae divided and shared her holdings with her two oldest sons, M. S. Bringier's worth at the Houmas property totaled $500,000 and Amedee Bringier's assets at the Hermitage amounted to $170,000.[181]

During the ten-year period from 1851 to 1861 culminating with the Union occupation of the river region during the Civil War, the Bringiers produced an average annual crop of 550 hogsheads at the Hermitage

Land and Livelihood 69

and 1,555 hogsheads at their Houmas/Brulé property.[182] However, in spite of the profitable seasons noted, there was no guarantee of success in the production of sugar. In addition to problems associated with market variables and management decisions, the crop was susceptible to plant diseases, rain totals, floods, hurricanes, freezing temperatures, and manufacturing and labor difficulties. These factors could lead to noteworthy differences in the size of an estate's crop. As demonstrated in the following table of crop production figures of the Bringiers' three sugar producing units, depending upon the variables noted above, production totals on the family's lands from one year to another showed little consistency.

Antebellum Sugar Production Totals and Production Output of Bringier Properties[183]

Year	Hermitage		Houmas		White Hall	
	Number of Hundreds	Percent of Change	Number of Hundreds	Percent of Change	Number of Hundreds	Percent of Change
1844	505		1,170			
1845	660	+31	966	-17		
1846						
1847						
1848						
1849-1850	380		995		251	
1850-1851	350	-8	950	-5	46	-82
1851-1852	496	+42	607	+36	298	+548
1852-1853	560	+13	1,926	+217	153	-49
1853-1854	662	+18	2,400	+25		
1854-1855	530	-20	2,100	-13		
1855-1856	395	-25	1,100	-48		
1856-1857	145	-63	675	-39		
1857-1858	900	+521	1,140	+69		
1858-1859	390	-57	2,155	+89		
1859-1860	280	-28	1,350	-37		
1860-1861	385	+38	700	-48		
1861-1862	1,250	+225	2,000	+186	75	

See Appendix C for additional crop figures.

Although freezing temperatures, such as those that afflicted the area in early 1852, could wreak havoc on an estate's sugar production, sugar growers were grateful that severe winter temperatures only threatened the area on an irregular basis. It was the summer storms and more frequently the overflowing of the region's waterways, particularly the Mississippi River, which produced the most persistent of Mother Nature's threats to the sugar crop and the plantation. South Louisiana's system of rudimentary levees provided only limited protection from the floods, which all too frequently visited the area around the Bringiers' properties. The need to keep the fields properly drained during rainy periods—so that setting water would not damage the moisture-sensitive crop—was a planter's major concern. However, it was the threat of a crevasse or broken levee which most worried the sugar growers along the river. A major break in the levee anywhere in the vicinity of a plantation during a period of high water could result in the inundation of a planter's property and spell ruin for a crop. For members of the Bringier family the period between 1849 and 1851 marked the time when their estates were most threatened by flooding. During the 1849-1850 growing season floodwaters inundated and destroyed a large portion of the crop at the Hermitage. Unfortunately, the next year the situation threatened even worse damage when between fifteen and eighteen crevasses fractured the levee system between New Orleans and the spot north of Baton Rouge where the Red River intersected with the Mississippi. The overflow situation was so bad in the Ascension Parish area that several of the largest planters in the region, including Duncan Kenner at his Ashland plantation, suffered huge crop losses. Even when their own lands were spared from the flooding, the periods of high-water threats cost planters considerable effort and expense. Upkeep of the levees, including the emergency repair of crevasse damage, was largely the responsibility of the planters and their workforces. Though a particular crevasse may not have been a direct threat to a planter's own estate, when a break threatened anywhere in the region, the Bringiers and their neighbors pulled their workers from their normal duties to put them to work on levee repair and maintenance.[184]

In spite of the danger that bad weather brought to the industry, staple-crop agriculture over the long run was profitable for both the large-scale sugar and cotton growers in Louisiana. Nevertheless, they were not immune from the market and economic variables, which negatively affected the economies of the nation and state during the antebellum period. To protect against financial disaster by providing additional revenue sources, planters sometimes diversified some of their crop production and invested in other business endeavors. Doradou's diversification efforts provided him with financial resources that the Bringiers were able to rely upon when their sugar business was underperforming. In addition to successful real estate activities in New Orleans, Doradou grew tobacco as a secondary money crop. A distinctive type of tobacco called perique exclusively grew in a relatively small region adjacent to the Mississippi River in a narrow corridor in the vicinity of St. James Parish. Although its distinct flavor made it a sought after commodity, efforts to raise the unique tobacco elsewhere failed.[185] Doradou realized that the crop's limited availability offered a great business opportunity and profit potential. Fortunately, the family's Brulé estate—located in the rear of their Houmas property—was an area that contained the unique soil and climate conditions needed to grow the treasured perique tobacco. Doradou dedicated a large segment of Brulé to the growing of the plant. Once the crop matured, it was harvested and moved to the Hermitage where it was processed by placing the stripped tobacco leaves under pressure in barrels and stored for several months to ferment and obtain its characteristic robust and pungent flavor. Doradou acquired considerable celebrity and won acclaim for his special crop. Even European agents sought his tobacco. After Doradou's death, great care was taken by the Bringiers to protect the quality and value of the crop. Only small amounts of it were sold at any one time, usually through special agents including the Baring Brothers of London who acted as the European handlers of the valuable product. Stored in a special brick building on the estate, an inventory taken shortly before the Civil War placed a value of $180,000 on the stock of tobacco stored at the Hermitage.[186]

Although many factors influenced the success or failure of a sugar plantation, an estate's profitability was primarily dependent upon the skillful management and direction of the planter. During the decades that the family worked the fertile alluvial soils along the banks of the Mississippi River, demand and mostly profitable prices for their crops—coupled with talented management—provided the Bringiers with a lifestyle that only a comparatively small number of Southerners enjoyed. With the combined improved acreage of 2,690 acres and unimproved acreage of 44,140 acres, the Bringiers' Hermitage and Houmas/Brulé plantations were considerably larger than most of the other large slaveholding sugar estates in Ascension Parish, which averaged 1,174 approved and 3,017 unimproved acres. Similarly, the 530 slaves held on the two estates far surpassed the average of 186 slaves found on the other large slaveholding plantations of the area.[187]

As to the question of the overall value and worth of the Bringiers at the end of the antebellum period and before the onset of their financial decline, which the Civil War brought to Louisiana, the answer is not clear. Census reports that list personal and real property values for the family do not address the full extent of the family's worth. For example, based upon parish-specific calculations only, the 1860 count shows the value of Amedee's holdings at the Hermitage as $80,000 in real property and $90,000 in personal property and the value of M. S.'s real property at $250,000 and personal property at $250,000. These figures appear to exclude Aglae Bringier's share of the estates and definitely fail to include her Melpomene home and other real estate interests in New Orleans. The figures cited for the Hermitage also fail to take into consideration the fortune of perique tobacco stored on the property. It is clear that through the colonial and antebellum periods the family's plantations provided the Bringiers with great wealth and status and were among the largest and most profitable to be found in Louisiana.[188]

Chapter 7

Masters and Slaves

The Bringier family spent generations working to establish and successfully operate their plantations along the Mississippi River. They were not alone in their endeavor. Without the effort and labor of many others who over the years assisted the family as either paid or forced labor, the Hermitage, Houmas, and their sister estates of the Bringiers could never have risen from the overgrown and forested alluvial lands of southeast Louisiana. The imposing houses the Bringiers and their fellow planters so loved would have remained unbuilt and an inestimable number of acres of sugar cane and other agricultural products would have remained unplanted, unattended, and unharvested without the sweat, toil, and skills of numerous workers and other non-family members. Despite the vast difference in power and social standing between the planter and his workers, most of whom were slaves who had little or no choice about their place and conditions of work, all who lived or worked on a sugar estate shared the common fact that their lives evolved around the production of a sugar crop. To produce the quantity of sugar that the Bringiers achieved on their plantations required sophisticated agricultural and mechanical facilities, which in turn necessitated not only a large workforce but also a sizeable number of workers and artisans, including whites, slaves, and free persons of color, who possessed specific skills and talents.[189]

Thanks to the post-emancipation memoir of one of the Bringiers' most accomplished and favored slaves, Pierre Caliste Landry, we have a listing of many of the workers and artisans used by the family on their Houmas plantation and most probably, since the Bringiers frequently shifted workers and equipment from one plantation to another, at the Hermitage and other family properties. Among the artisans who worked on the Bringier plantations were the following individuals: Ursin Boudreaux, general superintendent of mechanical and physical laboratory and shops; James Lea, chief engineer; James Malone, in charge of cooperage; Nicholas Allen, head blacksmith; Ernest Alexie, copper and tinsmith; a French-educated free person of color, Eugene Roudenez,[190] master bricklayer; Pierre Carmouche, bricklayer foreman; Dr. Garbriel, veterinarian, horticulturist, and floriculturist; and Professor Thomas

Lardner, director of truck farming, grafting, tree budding, and landscape gardening. Medical services were provided by Dr. Joseph Cottman and his cousin Dr. Thomas Cottman. They were assisted by "Aunt" Harriet Carroll who served as the hospital matron and surgeon's assistant. Other staff who were primarily assigned to the Houmas estate included Joseph Burbridge, the chief butler; William Oliver, head cook; and George Hampton, coachman and boy servant. Women also assisted in directing operations on the plantation. Under the supervision of Mrs. Bringier, "Aunt" Emma was in charge of what was called "plain" sewing and "Aunt" Ellen directed the fancy needlework and was in charge of the nursery. "Aunt" Jane supervised the laundry and "Aunt" Carmelite directed activities in the dairy.[191]

The Bringier family's association with slavery predated their arrival in Louisiana. When Marius Pons and his wife Francoise first arrived in Louisiana to establish their new home they were accompanied by slaves from their former home on the island of Martinique. From that time on, slave labor was an integral part of the family's story in their new land of settlement, growth, and prosperity. As is the regrettable case with the histories of so many African-American families and the stories of their ancestors, the names of most of the individuals who served the Bringiers throughout the colonial and antebellum periods and the specifics of their contributions to the Bringier's story are lost. Few records remain, other than some listings of first names, which provide few specific details of the hundreds of men, women, and children who labored over the decades of the late eighteenth and more than half of the nineteenth century on the Bringiers' plantations. Forced to perform the vast majority of the manual labor performed on the family's properties, their sweat and toil made the Bringier family's estates among the most productive and profitable plantations in Louisiana. Their contributions to the Bringiers and society as a whole must be judged by examining the results of their labor, which included a work product that kept the Bringier's crop production totals among the highest and most profitable in the state and the construction and maintenance of the family's impressive collection of manor homes that populated the River Road in St. James and Ascension parishes.

From existing records and documents, there is no evidence to suggest that the Bringiers' attitude toward slavery differed in any substantial way from the majority of their white Southern counterparts. As noted by historian William Scarborough in his work on the elite shareholders of the Old South *Masters of the Big House*, the vast majority of both male and female members of the elite slaveholding class in the antebellum South "had absolutely no compunction about their ownership of human property."[192] Since colonial times in both the North and South, racist sentiment was widespread among white Americans, even including many of the Northerners who favored the abolition of the institution. Nevertheless, Southerners carried the distorted notion of the black man's racial inferiority much further than their northern neighbors. They used the alleged lowly status of the Negro to justify their absolute denial of freedom of the bondsmen of African descent who populated the South's plantations. Although deep-felt racism formed the core of their defense of slavery, Southern whites offered a hodgepodge of other pro-slavery arguments to buttress their defense of the demeaning institution. Among their critical pro-slavery arguments was the practical justification that the enslavement of millions of blacks was essential for the continued prosperity of the South and ultimately that of the entire United States. Southerners were convinced that slavery was the necessary component that made possible the cotton industry that propelled America's economic growth during much of the first half of the nineteenth century. Louisianians were equally certain that it was the peculiar institution that drove the sugar industry, which was the foundation of much of antebellum South Louisiana's wealth and prosperity. In a less practical sense, but just as deeply believed in and promoted by slavery's defenders, was the belief that Scripture unquestionably justified the institution. Because the Bible contained no explicit directive against it, Southerners fervently maintained that slavery was a divinely sanctioned institution. Finally, in conjunction with these arguments and in spite of the increasingly bitter criticisms of Northern abolitionists, white Southerners insisted that slavery in the South was morally righteous and that it produced positive benefits for everyone involved in the institution, including the slave, master, and society as a whole. In support of this position, Southerners boldly argued that instead of being an oppressed people, the black bondsmen of the South were wards of loving, paternalistic masters who provided their slaves with an unparalleled level of care and protection which provided their chattels with a standard of living superior to the supposedly free workers in the North and in Europe.[193]

Despite the large disparity in power and in the social, economic, and political standing between them, slave and master on a sugar estate shared a commonality of dependence upon the success or failure of the estate's money crop. To successfully produce the crop year after year required the coordinated efforts of everyone who lived on a plantation. This need for coordination of effort influenced the organization of labor and responsibilities on the estate. Sugar making was too complex a process and the needed workforce was much too large to produce sugar on plantations the size of the Hermitage and the family's other sugar producing properties for one individual to have directed it alone. On the Bringier's plantations, as on nearly all sugar plantations of any size in the South, there developed a hierarchical labor organization pattern, which harmonized with the needs of the estate's sugar production. At the top of the work pyramid was the planter who, regardless of whether they originally obtained their wealth through inheritance, good fortune, or hard work, in order to sustain his position and standard of living had to possess no small amount of energy, perseverance, and managerial skills. However, even the most talented of planters relied on several functionaries to assist them in the successful operation of their estates. The most prominent of these was the planter's factor or commission merchant. As noted previously, the Bringiers relied on close friends or family members to provide this service. Most notable of the family's factors was Doradou and Aglae's son-in-law Martin Gordon, Jr. Usually working for a commission of 2.5 percent of the gross proceeds, factors were essential for the financing and marketing of the plantation's all important commercial crops. With planters spending most of their time on their estates in the country, factors included among their duties the supplying of plantation supplies, personnel items, and travel arrangements for the family.[194]

On the plantation itself, the overseer provided the planter with the most valuable assistance. Usually filling the role of day-to-day manager of the estate's operations, overseers carried out the general policies as determined by the master and played an essential role in the successful operation of most large properties. Because of the sophistication of the sugar-making process, overseers on sugar estates required a higher degree of technical knowledge than their counterparts on cotton plantations. The need for some technical skills and the fact that sugar estates generally had larger slave forces to manage than cotton estates resulted in sugar overseers usually being paid higher salaries than their peers working on cotton plantations. The job was a difficult one. In addition to the technical issues involved in producing a sugar crop, the overseer was the man in the middle of the master-slave relationship. Planters, at the level of the Bringiers, maintained a tenuous balance between their self-envisioned roles as patriarch of their black slaves and entrepreneur seeking to maximize crop production and profits. Hence, the overseer had to somehow meet both expectations of his boss—not overworking the slaves, which could possibly cause unrest among them, and, at the same time, getting the estate's workforce to produce the ever-desired bountiful crop. Considering all of the variables that came into play in the operation of an estate, the magnitude and sophistication of the overseer's task was such that job security for the profession, which was considerably lower on the social scale than that of even a modest planter, was typically not great.[195]

On some larger estates more than one overseer was used. On Doradou and Aglae's plantations, their older sons also provided day-to-day management assistance. Planters' sons often became overseers so that they could be better prepared for later careers as planters. Although sons of small planters made up a significant number of the overseers who worked in South Louisiana's sugar region, the sons of wealthier planters, including the Bringiers, most often spent their time learning the estate's operations by assisting in the management of their family's assets. Marius Pons followed this practice by involving his sons Louis and Doradou in the management and operations of his White Hall plantation and other properties. The practice was continued years later when Doradou and later Aglae relied upon their sons, M. S. and Amedee, to assist in directing the operations of the family's plantations.[196]

The estate's slave drivers also assisted in the management of the plantation's workforce and crop production. The master or his overseer usually chose these members of the slave force for their loyalty, physical strength, leadership ability, and dependability. They held special position among the slave population of the manor. Because the tenure of overseers was often not long, the drivers of the plantation provided a measure of stability to the estate's operations. Typically, they assisted the overseer in directing the work of the slave gangs that carried out the manual labor tasks on the plantation. The slaves of an estate were usually divided into groups or gangs based upon sex, age, and ability. Under the gang system, the plantation's field hands worked in groups under the direction of a driver for a specified period of time each day. Gangs with the stronger members were usually assigned the more strenuous tasks such as plowing and breaking up the soil for cultivation. "Hoe gangs" were assigned the less physical, yet just as unrelenting, maintenance tasks such as keeping the weeds down in the fields and the cleaning of drainage ditches. Of course, during harvest or rolling season all hands were involved in the sugar-making process. Although the gang or timework system predominated on most Old South estates, the diverse nature of plantation life sometimes led planters on occasion to resort to the use of the task or piecework system to assign jobs to their slaves. This usually occurred when individual workers were given specific duties to perform. This work system was commonly used in directing the work of the more skilled slaves such as carpenters, coopers, and smiths and when assigning easily measured tasks such as wood cutting and ditch digging.[197]

The cultivation of sugarcane and its processing into sugar was always a labor-intensive undertaking. Planters persistently strove to maximize their worker's productivity. The Bringiers and others realized that force and intimidation alone would not produce maximum work and good behavior among their slave workers. To entice more production from their slaves, some planters—including the Bringiers—provided incentives such as tobacco rations and other items as rewards for improved work output. However, historian J. Carlye Sitterson—in his notable study of the sugar industry, *Sugar Country*—credited the Bringiers for having one of the most interesting and extensive slave incentive systems among all sugar plantations. According to Sitterson, the Bringiers operated a plantation store, which they stocked with many articles favored by their black charges. Among the most popular items offered were meats, flour, shoes, calico, tobacco, and handkerchiefs. A special ledger was kept that indicated in one column purchases as debts and in another work output as credit for such activities as chopping wood, making the large sugar casks called hogsheads, and raising corn. Female slaves received credit at the rate of fifty cents per day for their labor. The special accounts were so popular that the ledger shows that credit balances often totaled $50 and in a few rare cases as high as $100. Most slaves on the plantation were allotted one pint of molasses a day by the overseer, which they could use to barter for items offered at the store.[198]

On the family's Houmas estate, the store was operated, with the permission of the Bringiers, as an entrpreneurial enterprise of two of the family's favored and most enterprising slaves, Joseph Burbridge, the household's chief butler, and a young slave Pierre Caliste Landry. Little is known of Burbridge except that he was chief butler in the family household and as such enjoyed a degree of privilege among the plantation's slaves. Caliste was a young mulatto purchased by M. S. Bringier for $1,665 in 1854. He had grown up on the neighboring plantation under Dr. Francois Marie Prevost's lenient ownership and was acquired by the Bringiers when the doctor's estate was auctioned off following his death. At the time, Caliste's mother—a free person of color—was placed under the guardianship of M. S. Bringier's brother-in-law, Martin Gordon, Jr. Although the Bringiers punished other slaves for being able to read and write and for giving instruction to other bondsmen, Caliste, who knew how to read and write at the time of his purchase, was not punished and was only given the warning by his mistress that "we . . . know that you are a trust-worthy and well-disposed servant and we are willing that you should learn, but you must not try to teach the plantation laborers."[199] The Bringier's favorable disposition for Caliste was further demonstrated when upon his arrival at Houmas after his purchase, he was installed as chief pantryman in the big house where he had the responsibility of sleeping in a room adjacent to the "laboratory and nursery" so he could answer the night calls of family members. Later he was given the job of "superintendent of the yard" where he directed the activities of the other yard servants who were responsible for much of the plantation's domestic affairs and care of the young white children on the estate during daily outings, such as horseback riding, fishing, and game playing.[200]

The operation of the little store, which the proprietors proudly called the firm of "Joe and Caliste," was so successful that M. S. Bringier and his overseer allowed the enterprise to expand to include a moss-press, broom factory, and wood yard. Most unique of all, was the establishment of a contract service with the overseer in which the never-ending and time-consuming task on the plantation of ditching was contracted to the "Joe and Caliste" firm who in turn sublet, apparently on a volunteer basis, to plantation hands the odious job of keeping the estate's ditches functioning properly. "Joe and Caliste" remained in operation until 1862 when the two partners dissolved the operation by mutual consent.[201]

Although innovative, the plantation store and subcontracting effort were not the only example of the Bringiers' efforts to provide their slaves incentives to keep them happy and productive. Sometimes they simply used cash to reward the work of their slaves, particularly when there was a bountiful harvest. This was the case in 1852 when the family's factor and Melpomene resident Martin Gordon, Jr., boasted that "not another plantation in Louisiana" could match the sugar production of the family's Houmas plantation and suggested to his in-laws that the workers be rewarded with silver payments because, "The boys have no doubt worked well . . . and a little money given to them, would do good."[202]

Another common inducement practiced by the region's planters was to encourage their charges to cultivate crops and raise poultry, rabbits, and hogs on their own as a means of supplementing the supplies provided to them by their masters. This allowed slave families to earn small amounts of spending money and, in the case of the Bringier family slaves, to earn credits to be used at the plantation store. Although records are scarce on the issue, the fact that slaves on the Bringiers' plantations were able to earn modest cash rewards to spend at the store makes it likely that the Bringiers also followed the common practice of providing their slaves with small patches of land for their personal use.[203] Additionally, on the plantations of Doradou and Aglae's daughters, Marie Elizabeth "Algae" Tureaud and Nanine Kenner, slave incentive programs were in place. Duncan and Nanine Kenner had the more ambitious of these efforts at their Ashland estate where the slaves raised large numbers of chickens on their personal plots. However, Duncan, ever the master and businessman looking for a means to earn revenue from his workers' efforts and as a means of keeping track of their wealth, required his slaves to sell their poultry directly to him at the rate of twenty cents a pair, which he promptly resold for thirty cents. However, Kenner's hand in his slaves' chicken concerns did not stop them from accumulating some cash and status on the plantation. For example, other slaves at Ashland spoke about Old Cudjo in near legendary terms as he successfully amassed a slave's fortune of more than 500 silver dollars.[204]

Of all of the incentives and awards dispensed to the slaves, the activity Doradou and the other members of the family enjoyed the most was an annual tradition of treating the plantation's younger slaves to a special favor at Christmas. Standing on the Hermitage's back gallery, Doradou gathered the plantation's young blacks in the yard below where they awaited the distribution of their annual Christmas gifts. Then Doradou would throw handfuls of coins to the young slaves assembled where they scrambled and competed for the coins in a fashion not unlike children scampering for throws at a New Orleans' Mardi Gras parade.[205]

The slaves of the Bringier and other planters across the South were dependent on their master's generosity for more than just the petty incentives offered to promote productivity. Their masters determined nearly all of the conditions and circumstances under which the slaves of a plantation lived and labored. It was the master who provided the food, clothing, housing, and health needs of the plantation's slave force. Anxious to keep their labor force as viable and productive as possible, planters generally provided their slaves with a reasonably adequate supply of food. Although plentiful in quantity, the slaves' diet lacked variety and nutritional composition. However, the nutritional deficiencies of the diet were not the result of planter neglect. Rather, the dietary problems of slaves were more the result of the state of antebellum medical knowledge than the condition of slavery. Little was known at the time about vitamins and nutritional needs. Indeed, both white and black Southerners of the antebellum period consumed diets that were nutritionally unbalanced. Over the years, the diets of both black and white Southerners gradually improved. By the early nineteenth century, it was a widely accepted standard across the South that the average ration for healthy adult field hands was approximately three and one-half pounds of pork or bacon and eight quarts of cornmeal per week for each slave. Typically on sugar plantations there was an additional ration of molasses provided. Archaeological evidence from plantations in the area around the Hermitage reveal that the region's planters also supplemented their slaves' diets with portions of beef—typically taken from the less desirable parts of the animal including the head and less meaty parts of the body—fish, and wild game which abounded in the area. Raccoon, opossum, rabbit, wild birds, catfish, freshwater drum, gar, sunfish, and mackerel all supplemented the mundane diets of southeast Louisiana's slaves.[206]

With most of the food needed being produced on the estate, the cost to planters for feeding their slaves was minimal. Providing clothing for their slaves was a different case and often required planters to pay substantial funds to outfit their workers. Because of their greater financial resources, the masters of larger plantations were in a position to be more generous in dispensing clothing than their lessor slaveholding neighbors. Slaves' clothes were fairly crude but functional. Typically on large estates slave clothing was distributed twice a year in the spring and fall. Field hands commonly received four suits of clothes each year with males being outfitted with pants, shirts, jackets, hats, handkerchiefs, and shoes. Women were provided dresses and frock-type garments called joseys and young children received long shirts. On most large plantations, the majority of the bondsmen's clothing was made by the estate's women slaves, who were provided the required materials for this task by their masters who obtained the cloth in bulk purchases from outside providers. For example, in 1852, the Bringiers paid their factor in New Orleans $1,350 for a single shipment of material destined for use on the family's plantations to provide winter clothing for the male field hands. Items which could not easily be manufactured on site—hats, jackets, blankets, and shoes—were purchased as finished products. In September of 1845 in preparing for the upcoming winter, Doradou paid $175.85 for a bale of blankets for use at the Hermitage and $350.25 for two bales for use at his Houmas plantation. The very next year he again needed blankets for the large slave force at his Houmas property and purchased twenty additional pairs for $70.50. In addition to blankets, shoes were among the most costly items furnished to slaves. As with the purchase of blankets, Doradou preferred to replenish his supply of shoes for his bondsmen as part of his winter preparation activities. This had the advantage of having new clothing and personal articles on hand for his workers prior to the worst weather months of the year. In 1846 Doradou made two large purchases of shoes for his workers, including an order in October for thirty dozen pairs at a cost of $200.25 and a smaller acquisition of thirty-six pairs in December for $36.25. Buying in bulk permitted Doradou to obtain his shoes at a cost of only $.50 cents to a $1.00 a pair. These rates were considerably cheaper than the $1.00 to $1.25 a pair that became the going rate by the later years of the antebellum period. No comment is recorded in the Bringiers' plantation ledger about the sizes or quality of his shoe purchases. Because of the poor quality that was typical of shoes purchased by planters for their slaves, no other item of clothing was more complained about by the bondsmen than shoes. One way to understand the quality of slave footwear is a comparison to those worn by their masters. In the same ledger where he recorded Doradou's shoe purchases for his slaves, factor and son-in-law Martin Gordon, Jr., also noted new shoe acquisitions for the Bringiers' of two pairs for $5.00 and $8.00 for Aglae and a single pair for $4.00 for Doradou. Nevertheless, whether it was the questionable quality of the footwear or from habit and preference, during most of the year large numbers of slaves worked barefoot in all but the worst weather.[207]

Besides clothing and feeding them, planters also had the obligation of providing adequate housing for their workers. During the colonial period, slave housing, like that of many poor whites of the time, was fairly primitive. As the sugar industry took hold in South Louisiana during the early nineteenth century and the number of slaves working on large estates increased, greater resources were committed to the housing for slaves. Beginning in the late 1830s and continuing until the war, southern agricultural journals promoted efforts to improve the quality of slave housing as an important factor for planters to consider in protecting their financial interests. Good housing promoted good health,

Masters and Slaves

which buttressed Southern claims during the slave debate of paternal treatment of their slaves. Better housing also had the positive financial effect of doing nothing to limit the reproductive activities of the estate's slave population.[208]

Specific details about slave housing on the Bringier plantations are limited. In a plantation journal that included information for several years of plantation data ending in 1847, the year he died, Doradou noted that at his Hermitage plantation the slaves were housed in ten large cabins that measured 30 feet square. The slave force at the Hermitage, who were housed in the ten cabins mentioned by Doradou, totaled sixty adults and twenty children. If accurate, at 900 square feet, the cabins were several times larger than the more-or-less average size slave cabin found on most plantations of the time of 300 to 400 square feet, which were usually laid out in approximately 16 by 18-foot configurations. Most slave shanties were raised a few inches off of the ground and were of wood-frame construction. Styled somewhat similarly to the houses used by poor whites of the time, the structures showed few direct African influences. Usually built as two rooms with a fireplace between the rooms, the sizes of the Bringier cabins suggest that the structures were arranged in the less common, but not all that unusual, four-room setting. Not unexpectedly, as they increased the crop acreage and production on their properties, the Bringiers concurrently increased the number of slaves and their housing on their plantations. By 1840 the Hermitage's slave force totaled 116 and was housed in twelve of the larger cabins noted above. By the time Aglae divided ownership of the Hermitage with her son Louis Amedee in 1858, the number of slaves on the estate had grown to 157. Two years later, however, the number of slaves on the estate dropped by thirteen to 144. Existing records do not reveal the reason for the reduction in the slave population. It is possible that several could have died or had been sold during the intervening period or that they were simply reallocated to another of the family's properties at the time the count was taken. On occasion the Bringiers moved slaves from one of their properties to another. With the Houmas property and its adjoining Brulé estate traditionally being the Bringiers' prime crop production site, it is possible that the missing Hermitage workers were relocated there. The same census that showed the Hermitage with 144 workers revealed that the Houmas plantation workforce was more than double that number with 386 slaves.[209]

As noted previously, the slave-cabin configuration used at the Hermitage included larger-style buildings, which resulted at times in a greater number of slaves being housed in a structure than was common on most plantations toward the end of the antebellum period. By the end of the last decade before the Civil War, the average number of persons housed in slave cabins on large estates in the South was between five and six individuals—or one family unit.[210] The common effort in the South to improve the quality of housing for the region's slaves was not ignored by the Bringiers. In 1833 at the Hermitage, each slave residence averaged eight persons. By 1840 the number per slave household on the estate grew to nearly ten. However, twenty years later with the plantation's 144 slaves housed in twenty-six structures, the number decreased to 5.5 per house, a number only slightly higher than the southern average for the time of 5.2. Indeed, at the Houmas plantation the 386 bondsmen were quartered in ninety-six buildings with a ratio slightly greater than four per house—a rate considerably lower than the average slave occupancy rate found on most large contemporary plantations.[211]

The sugar industry's rapid expansion across southeast Louisiana resulted in a moderately steady increase in the demand for slaves. As was typical among regional sugar planters throughout the colonial and antebellum periods, the Bringiers were nearly always in need of additional laborers for their plantations. To meet the demand for slaves on his estates, Doradou often used his father-in-law's firm of DuBourg & Baron to locate and purchase slaves for both his city and country properties. For example, in 1817 the DuBourg & Baron firm paid $1,200 to the slave brokers Hubert & Brown for two slaves for Doradou. Because there was rarely a surplus of slaves in the New Orleans market, the Bringiers and other planters at times imported slaves from out of state to meet their workforce needs. This was the case in late 1827 when Doradou, again working through his father-in-law's company, acquired thirty-four slaves from Albert Gallatin Cage of Mississippi for $15,500.[212]

Not unexpectedly, the value of the Bringiers' bondsmen varied over the decades that their plantations were worked by gangs of forced laborers. One way a planter could leverage the cost of new slaves was to make a group purchase of new bondsmen. This was the situation with the 1827 purchase by Doradou of thirty-four slaves noted above. By paying a single price of $15,500 for the group, he paid an average price of only $456 for each of the slaves. The going rate for a prime field hand at the time in south Louisiana's sugar-country was roughly $600, saving Doradou approximately 24 percent on his purchase. Another major factor that influenced the cost of slaves was the country's economic condition. For example, leading up to the nation's economic collapse of 1837, the prices for prime male slaves averaged $1,300. After the crash, the value of slaves decreased to such a degree that two years later the average price for a male field hand in the sugar-growing regions of the country was only $800. As the American economy recovered and as the sugar industry expanded its production, the prices for slaves rebounded. By May 1844, M. S. Bringier paid $900 to acquire a "Negro boy." Prices continued to experience a steady increase until the eve of the Civil War when a skilled bondsman sold for as much as $3,000.[213]

Slaves also possessed unique characteristics, which either added or subtracted from their value. The age, size, skills, health, gender, and demeanor all influenced the value of a slave. For the most part, male slaves typically were worth more than females of similar age, skilled workers were more valuable than simple field hands, and younger slaves were worth more than the elderly. As noted in Appendix D, the individual value of the Bringiers' slave force at the Hermitage shortly before the war varied considerably. An 1858 inventory by Aglae and Louis Amedee Bringier included a list of the value of the plantation's slaves. Taken as a whole, the 158 men, women, and children were valued at $107,700, for an average worth of $682 each. However, the average value cited represented a wide range in the individual worth of the members of the Hermitage slave force. For example, Mary Norman, identified as "crazy," was listed in the inventory as having no value and forty-year-old Christmas, who served as the plantation's blacksmith, was listed as the Hermitage's most valuable individual slave at $1,800.[214] Other notable members of the workforce included the estate's two engineers—George, at age twenty-five was valued at $1,600 and forty-eight-year-old Jesse at $1,400. Also listed as workers with special skills were the plantation's coopers—William whose worth at age forty-two was noted as $1,500 and seventy-two-year-old Simon who served as the plantation's carpenter (due to his advanced age he was valued at $700). Most of the able-bodied males of the Hermitage's slave force who were not listed as having a special skill or who were not suffering from some illness or other problems were generally valued at $1,400 each.[215]

The value of a slave usually decreased as they aged. The treatment of superannuated slaves was a topic of concern among Southern planters. As a group, planters in their debate with the anti-slave critics in the North boasted of their humane treatment of their aged workers as an example of the paternalistic nature of the peculiar institution. However, research by modern historians has revealed numerous examples of planters' indifferent or even hostile treatment of older slaves who had grown to be financial liabilities to their masters. Nevertheless, existing records suggest that the Bringiers were among the planters who counseled full and kind concern for their senior and disabled bondsmen. Unfortunately, for those individuals, whether white or black, born to the lower classes of society in America during the antebellum era, the combination of poor living standards, heavy physical labor, inadequate medical care, and cheap housing all resulted in a high mortality rate, which led to relatively few members of the lower classes in the Old South reaching their seventies. This deadly fact was as true for the Bringier bondsmen as it was for most poor Southerners.[216] At the time, approximately 40 percent of slaves died before reaching age nineteen and in the South both blacks and whites alike were considered "old" at age fifty. At the Hermitage in 1858, 6 percent of the slave force was older than fifty-nine, with four individuals or 2.6 percent of the slave force being seventy or older. As slaves reached the point when they were too old to perform the physically demanding work in the fields, they were typically given less-strenuous jobs to perform such as feeding and caring for the plantation's animals. No longer able to perform the primary task of a plantation slave—heavy-duty physical labor in the fields—the value of the aged bondsmen quickly declined. Hence, among the Hermitage's labor force seventy-year-old Joe Congo's value in 1858 was listed as $200 and seventy-year-old Old Tom's worth was set at only $150. Sixty-five-year-old Bob, who possessed more skills and abilities than Joe Congo and Old Tom, was considered to have greater worth and was valued at $300. The most senior slave on the plantation was Tommy who, in spite of being crippled, at age seventy-six maintained a value of $100.[217]

Although in the South the number of male and female slaves above the age of fifteen was approximately equal (99.1 females per 100 males), of the slaves in that same age bracket who resided at the Bringiers' Hermitage plantation in 1858, the gender make up was somewhat different with 55 percent (fifty-nine) being males and 45 percent (forty-eight) females. Of the members of the plantation's slave workforce who were between the ages of twenty and forty-nine years old, males made up 53 percent and females 47 percent of that portion of the plantation's slave population. However, once the slaves of the plantation reached their fifties the gender ratio changed dramatically. The 6 percent differential of the below fifty adult-age group jumped to a 25 percent difference (fifteen males to nine females) between the number of elder (above fifty) slaves who resided at the Hermitage. Hence, for the female slave population of the Bringier's plantation, life expectancy was even bleaker than that faced by their male counterparts.[218]

Because of their use as concubines and servants in Africa and the Near-East, in the Eastern Hemisphere female slaves were typically more highly prized than males. However, universally in the Western Hemisphere, because of their use as physical laborers, male slaves were in greater demand. In Louisiana, female slaves of equal age and health typically sold for less than their male counterparts. This was the situation at the Hermitage where the value of the plantation's female slaves ranged behind that of the men. For example, twenty-seven-year-old Louisa was valued in 1858 at $1,100, while twenty-seven-year-olds Alfred and Joseph were appraised by the Bringiers to be worth $1,400 apiece. Similarly, seventy-year-old Frosine was worth $50 at the same time that seventy-year-old Joe Congo was valued at $200.[219]

Besides the aging process, illness or injury lowered the value of both male and female slaves. The quality of medical care received by slaves on the South's plantations was primitive when compared to modern standards. However, the poor level of care was not the result of neglect on the part of masters or overseers, but from the crude state of medical knowledge and practices. Desiring to keep as many of their bondsmen as possible working in the fields and earning money, planters generally sought to provide their slaves the best medical care they could afford. To look after the health of their slaves, the Bringiers used a neighbor, Dr. Thomas Cottman and his cousin, Dr. Joseph Cottman. Records from the Hermitage and Houmas estates show that Doradou paid Thomas Cottman a sizable fee of $500 a year to provide medical care for his workforce. In his review of the issue, historian Peter Kolchin noted that in spite of the primitiveness of the care given to slaves on large plantations, such as the Hermitage, where hospitals were common, it generally exceeded the level of care received by most Southern whites of the time. Nevertheless, the hard physical work with dangerous tools and the archaic medical care took its toll on the plantation's workforce. When Aglae and Amedee Bringier inventoried the Hermitage plantation's slave force in 1858, of the 153 men, women, and children slaves held on the estate, 11 percent (seventeen slaves including ten women, five men, and two children) were incapacitated to such a degree by either illness or injury that they could not carry out their work responsibilities at a normal rate. The degree that these health problems affected the value of the slaves is illustrated by the following comparisons. Whereas a healthy forty-four-year-old Peter Jones was worth $1,100, Jean Louis who at age forty was "affected with hernia" and considered to be worth only $700 and forty-year-old Frank was valued at only $200 because of his being "consumptive." Similar depreciations affected the value of the Bringiers' female slaves. As cited earlier, Mary Norman suffered from such mental problems that she was listed on the plantation's inventory as "crazy" and was listed as having no worth. Likewise, as a result of her being "sickly," the value of fifty-two-year-old Ruthy was lowered to only $50, a rate only a fraction of what other fifty-plus female slaves at the plantation were valued at the time of the inventory.[220]

In addition to providing gender statistics about the estate's slaves, the 1858 inventory of the Hermitage slave force provides additional insight into the master-slave relationship on the plantation. The inventory lists the slaves in alphabetical order and by gender. Although mothers and their children were listed together, little information is provided on the makeup of slave families on the plantation. Because of the legal nature of the document, the inventory was no doubt prepared to conform to the legal standards of the day, which in regards to slave families largely ignored the institutional and legal support given to free families. Although marriages among slaves had no legal status, the majority of antebellum slaves lived in family groups. Despite being legal for a slave owner to separate spouses by sale, slave masters usually supported the establishment of slave families by recognizing the unions by such actions as making cabin assignments by family groupings. As is suggested by the grouping of mothers and children together on the 1858 inventory, some legal protection was given to the mother-child bond in the slave family, with mothers and their children less than ten years of age not being able to be sold separately. Whether because of the paternalistic feelings of their owners or the fact that the growing need for slaves made it impractical for planters to sell children once they reached the acceptable age for sale, during the antebellum era it was not common—though by no means impossible—for young children to be separated from their mothers.[221]

The Hermitage's slave inventory provides contradictory evidence as to the Bringier family's position on the buying and selling of slave children. Care was taken to list children with their mothers in spite of the fact that some children, such as Maria's sixteen-year-old son Juan, was older than fifteen and could have been included on the plantation's general list of workers. The grouping of the slaves in this manner demonstrates the Bringiers' commitment and respect for the mother-child union. On the other hand, the slave inventory of workers also included the names of seven individuals between eight and fifteen years of age who were not listed with a parent. Grouping these young slaves in the general list of plantation slaves instead of with the women slaves and their children implies that these seven children were not related to any of the plantation's adult females. Perhaps they were the orphan children of women slaves at the Hermitage or were slaves who had been acquired by the Bringiers on the open market. Although there is scant evidence to suggest which option applied, because of their young ages some of the younger slave children on the estate were likely orphans. Louisiana's *Code Noir* placed restrictions on the sale of young children including the prohibition of the separation by sale of children less than ten from their mothers. However, Amedee Bringier's comments in a letter to his wife Stella in early 1856 in which he related how he had advised his mother, Aglae, "to buy a young boy in the city" to serve as a new house servant at the family's Melpomene estate in New Orleans reveal that the planter and his mother had no objection to the controversial practice of acquiring children for their slave force.[222]

The Bringiers' untroubled decision to purchase the young slave boy exemplifies the contradictory, if not hypocritical, nature of the master-slave relationship on the plantation. Although most planters proudly defended the paternal characteristics of slavery on their estates, their concern for promoting the well being of their bondsmen sometimes clashed with their own economic or personal needs. Unlike interest conflicts in the family where parents often acquiesce at their own expense to their child's needs, on the plantation such paternalistic conflicts nearly always resolved to the master's advantage. Nowhere was the contradictory nature of plantation life more evident than the planter's position on the slave family. Though marriages between slaves had no legal status, planters generally found it to their advantage to support, to one degree or another, the existence of slave families. Despite the fact that the institution of slavery presented obstacles to the normal family relationships, the majority of plantation bondsmen in the South lived in slave families. The assignment of living quarters, the distribution of food and supplies, and the promotion of what historian Kolchin calls "family morality," which included the discouragement of immoral behavior, were all centered to varying degrees around the existence of the slave family. Yet it was the disruption, or more commonly the never-ending threat of the disruption, of the slave family that was one of the more heinous aspects of slavery on the plantation. Life on the Bringier plantations was not absent of the contradictions and issues noted. As was the case in 1794 when Marius Pons demonstrated some commitment and recognition of the importance of the slave family when he kept the family of eight-year-old Simon, his forty-three-year-old mother, and his thirty-two-year-old father together by selling the family as a single unit. However, there is little additional evidence that Marius Pons or other family members gave much notice or consideration to their slaves' relationship concerns. This was the situation evident in the 1858 inventory of the Hermitage's slave force where the Bringiers followed the common practice of recognizing the mother-child relationship while ignoring the fathers' participation in the family units of the plantation's slave population. The Bringiers' active participation in the slave trade by its very nature demonstrated a general disregard for the slave family unit. The buying and selling of slaves from places far or near commonly resulted in the disruption of family and personal relationship for the slave involved in a sale. [223]

A January 1856 letter from Amedee to his wife Stella further illustrates the contradictory and complex nature of the Bringiers' master-slave relationship. In the letter Amedee wrote of his intentions to send supplies and several slaves to the family to assist them while they stayed at Melpomene in New Orleans. Clearly demonstrating the extent of the power a master held over the lives and families of his slaves, Amedee noted that among the items being sent were "Emiline, Emile, some plants, ... eleven capons, and your mother's cook." It appears that by his inclusion of the transfer of the slaves in the same breath that he listed the capons and plants that Amedee gave little thought or concern to how the transfer would affect the lives and relationships of the slaves mentioned. Yet he went on to write that Stella needed to "tell mother that I did not send Antoine for when I spoke with Silvy of it, she came within a hair's breath of having a fit and she begged so hard to have him with her, that I made up my mind to advise mother to buy a young boy in the city"—a short statement that encapsulated the complex nature of the slave-master relationship. Though this statement revealed that the austere Amedee was willing to listen to Silvy's desperate appeal to not separate her from Antoine and could be cited as an example of master-slave paternalism, it more accurately underscores the basic heinousness of the slavery regime as practiced on the Bringier and every other plantation. Regardless of the benevolence of the master, day in and day out slaves' lives were based upon the master's power and whim and the absolute denial of freedom, which often made a slave's existence insufferable.[224]

Because of the scarcity of documents, few specifics are known of how the Bringier slaves behaved under their masters' tutelage. With such a large number of slaves and slave families present, it is likely that the demeanor of the Bringier slave community paralleled that of other large plantations across the South. The daily life of the slave was regimented. Although the Bringiers relied on incentives to encourage their workers to work hard, they no doubt used force or the threat of force when necessary as a motivator for obedience and adherence to the plantation's rules. Each planter and overseer developed their own system of punishments, which commonly consisted of extra or more distasteful work assignments. Serious problems were often penalized by forms of corporal punishment, with whipping being the most common form of physical punishment meted out against recalcitrant slaves. For those bondsmen who refused to conform to a plantation's regime, there was always the threat of sale to some far off place. Regardless of the rewards and punishments held over their heads by their masters, the burden of life under slavery on the Bringier plantations was more than some could tolerate. On most large and small antebellum plantations, it was not uncommon for unhappy slaves to address their despair by deserting the estate. Because of the distances involved and the difficulty runaway blacks had in surmounting the restrictions and controls placed upon their movements away from the plantation, relatively few slaves from the deep South successfully escaped to freedom in the North. Instead, the vast majority of truant slaves turned out to be only temporary fugitives who remained relatively close to their homes and who eventually returned to the plantation either on their own or by capture. Short-term truancy appears to have been the case in May 1852 when Duncan Kenner's overseer at his Ashland plantation noted in his journal that the slave George Brecks "ran off today."[225]

A more serious problem for planters to handle was those slaves who confronted masters and their assistants either directly or by sabotage of the plantation's equipment and structures. These actions resulted in planters applying severe punishments including harsh whippings, branding, cropping off of body parts, and execution. Such was the case in 1830 when Doradou posted a notice in the Baton Rouge *Gazette* warning of a runaway slave from the plantation named Isaac who according to Bringier was "Known as a very wicked fellow, capable of committing any crime." Before deserting the plantation, Isaac set fire to the estate's storehouse and in the subsequent confusion managed to steal $350 in cash and a supply of clothing. It is not known how the effort to apprehend Isaac and punish him for his actions turned out. However, from Isaac's physical description posted in the paper—one ear cropped and branding scars on his shoulder and on his chest—it is obvious that he had a long history of causing problems and for being severely punished for his untoward behavior.[226]

An even more serious attempt at slave sabotage occurred in 1854 at the Bowden plantation of Aglae's son-in-law Hore Browse Trist. According to Trist, a trusted slave named Old Pleasant

> . . . suffered the water in the boilers to get so low that there was scarcely any left in them, and when informed by some of the hands that there was something wrong, told them to mind their own business. The engineer . . . made his appearance about this time, and on going to the boilers found them heated almost to redness—He gave the alarm and bid all in the neighborhood run for their lives—but Pleasant instead of showing any concern . . . went and seated himself very coolly on one of the boilers . . . shortly after I [Trist] arrived another alarm was given; steam escaped with violence from the top of one of the boilers and made the ashes and brick fly. . . . This leak being stopped, steam was again raised when to our dismay it rushed out from other places and on cooling down and examining, we found several rents in the first boilers—it took five days to repair damages . . . but the syrup in the tanks and filters got sour and we made sugar of inferior quality for several days.[227]

From the time Marius Pons first arrived in Spanish-colonial Louisiana with the handful of slaves he brought with him from Martinique to the onset of the Civil War, the Bringiers worked tirelessly adding to the size of their slave force. Upon arriving in the gulf-coast colony, Marius Pons dedicated himself to the accumulation of wealth through dual aspirations of land ownership and commercial agriculture. The pursuit of these goals dominated life in Louisiana and the rest of the South for decades. More land required more labor, which led to greater crop production which resulted in greater profits and the improved lifestyle of the aristocratic planter elite. During the antebellum era, the growth of the sugar industry in Louisiana, its requirement of large numbers of laborers, and the family's growing reliance on the crop resulted in the gradual growth in the size of the Bringiers' slave force. In 1833 the workforce totaled sixty workers and twenty children on Doradou's Hermitage plantation. Seven years later, in 1840 the adult slave force on the estate nearly doubled to total 116 bondsmen. Even after Aglae assumed control of the Hermitage following her husband's death in 1847, the Bringier slave force continued to increase. The family's efforts to increase the size of their slave force were so aggressive that according to historian William Scarborough, by 1850 Aglae Bringier, with a workforce on her plantations of 673 bondsmen, held more slaves than any other female in America. Her slave holdings also ranked her second in Louisiana and thirteenth among all slaveholders in the South. Ten years later after she had entered into legal partnerships with and divided her various plantation and slave ownerships with her two sons, M. S. and Amedee, her slave holdings totaled 567. Although it is not clear as to why or how the number of slaves decreased on the Bringier estates over the ten-year period between census counts, the somewhat smaller number of slaves held either individually or in partnership by Aglae still ranked her among the largest slave masters in the country. Slipping in position to fifth in Louisiana, Aglae ranked thirty-first among the nation's nearly 385,000 slaveholders on the eve of the Civil War.[228]

Life on the antebellum plantation involved more than the home and business activities of the estate's master and his family. Though the planters were the driving force behind nearly all of the economic and cultural life of all who resided on the plantation, without the contributions of the estate's black residents the master's lifestyle and the essence of life in the antebellum South would have been radically different. As noted by Scarborough, "In the end, . . . despite the invaluable contributions of factors, overseers, drivers, and slave foremen to the success of planting endeavors, it was upon the onerous labor of the rank and file slaves that the planters depended for their prosperity, power, and prominence."[229]

Chapter 8

Conflict and Loss

The privileged lifestyle enjoyed and cherished by the Bringiers and other elite planters across southeast Louisiana came to a sudden and startling end in the spring of 1862. On the morning of April 24, alarm bells in the Crescent City rang out the fateful warning toll that the city's residents had long dreaded. The bells signaled the inevitable fall of New Orleans to the forces of the scorned enemy—the Federal naval force under Commodore David Farragut's command. The enemy flotilla had successfully fought its way past the two Confederate forts located to the south of the city on the Mississippi River which were supposed to protect the South's queen city from attack by Union naval forces. The unhappy event was a long time in coming. For decades Northerners and Southerners had debated the issues of slavery, states' rights, and the future of the Union only to have the dispute culminate in early 1861 with the secession of most of the slave-holding states and the formation of the Confederate States of America. In April 1861, Confederate forces in Charleston, South Carolina, opened fire on Fort Sumter, marking the beginning of the brutal and devastating war between the states of the North and the South and the beginning of the end of the Bringiers' cherished lifestyle.

Although there was a sizeable minority in Louisiana, including most of the sugar planters in the state who questioned the wisdom of secession and its potential for bringing commercial and social ruin to the South, among the members of the Bringier family there was near unanimous support for the Southern Cause. Only the sentiments of the husband of Aglae's daughter Louise, Martin Gordon, Jr., on the question of secession were much more ubiquitous than those of his in-laws. During the debate leading up to secession, he maintained a unionist position and advocated for the gradual emancipation of the South's slaves. Yet when Louisiana seceded he accepted the decision and offered his service to local officials. In the days leading up to the Union's capture of the city, New Orleans' mayor John T. Monroe appointed him to the city's Committee on Public Safety, which had the responsibility of coordinating the city's defense with both Confederate and state leaders. Nevertheless, when the city fell in April 1862, he remained a resident of New Orleans throughout the war. Whether the city was controlled by the Confederacy or the Union, Gordon managed to maintain access and good relations with the powers that were in charge. A leading merchant and sugar factor in the city, Gordon was adept at making the financial best of unpredictable situations. Even after the city's capture and in spite of the fact that he was registered as an enemy and that his in-laws were prominent members in the Confederacy, Gordon was able to establish close ties with the region's Yankee leadership. He was particularly close to Gen. Nathaniel P. Banks. The relationship between the region's Yankee commander and Gordon was close enough that Banks chose him as one of his emissaries to Richmond in June 1863 as part of an ill-conceived effort to negotiate an end to the war—a plan partially designed to catapult the Yankee politician and general to the forefront of potential candidates in the approaching 1864 presidential race. Although the scheme seems to have been given little interest by either the governments of Abraham Lincoln or Jefferson Davis, Gordon faithfully applied himself to the effort. Upon his return to New Orleans, he provided Banks with a report which included information examining the degree that the people of the Confederacy desired peace, an analysis of Secretary of State Judah Benjamin's efforts to win foreign intervention in the war, the potential of the South to arm its slaves to fight in the war, the status of the political and military fortunes of Jefferson Davis and Robert E. Lee, and the depreciated value of Confederate money. Banks was pleased with the effort and dutifully reported Gordon's observations to President Lincoln and praised him for "His standing and character . . . , and his opportunities for information from prominent public men of the South."[230]

Demonstrative of Gordon's ubiquitous loyalty, within weeks of his meeting with General Banks and reporting on his mission to Richmond, Gordon appeared behind Confederate lines in Shreveport where he met with Gen. E. Kirby Smith, the commander of forces in the Confederacy's Trans-Mississippi Department. In a January 1864 meeting with the rebel general, Gordon requested authority to obtain a supply of cotton for the stated reason of providing "relief of the suffering families of Confederate citizens and our soldiers there [New Orleans] held as prisoners of war."[231] Kirby forwarded the request to Gordon's brother-in-law Gen. Richard Taylor for review. It is not known if the request was granted. However, it is unlikely that Union officials would have simply allowed Gordon to bring rebel cotton through their lines to sell, even for "relief of the suffering families of Confederate citizens," unless there was some gain in the effort for themselves. It is unclear whether Gordon stayed in the area for an extended period or if he took advantage of his good ties with both sides in the conflict, which allowed him the unusual ability to pass through military lines almost unhindered. However, in mid-March he again met with General Smith. In this meeting, he provided the Confederate commander with an estimated size of Banks' pending Red River campaign invasion force. According to Smith, Gordon warned of "an overwhelming force" which so worried the rebel general that he fretted that his Confederate forces would be "forced back into the interior. . . ."[232] Unfortunately, the historical record does not provide the answer to Gordon's motive behind his visit to Confederate territory and his meetings with General Smith. Because of his close ties with both Union and Confederate leaders, his actions are subject to conflicting interpretations. Was his warning to Smith an effort to give the rebels a warning of the size of Banks' invasion force or was it a common wartime ploy to mislead rebel commanders causing them to retreat into the interior? What is known is that Gordon's visit to North Louisiana during early 1864 provided him the opportunity to obtain information about the region's valuable cotton crop. With one of the major objectives of Bank's Red River campaign of 1864 being the securing of desperately needed and valuable cotton for Northern textile mills, Gordon's travels to the area and his relationship with General Banks offered him a financial opportunity. Shortly after his meeting with Confederate General Smith in Shreveport, Gordon joined the other cotton speculators who tagged along on Bank's ill-fated Union campaign up the Red River in Louisiana. While on the campaign, he soon earned a reputation for trafficking cotton—this even though the Yankee invasion pitted a large Federal force under the command of his friend Nathaniel Banks against a much smaller rebel army under the leadership of his brother-in-law Richard Taylor. Banks' defeat and near disastrous retreat from the Red River region largely frustrated Gordon and the other cotton speculators' dreams of quick fortunes.[233]

Aware of the war's potential consequences, the majority of the Bringiers in the months leading up to the conflict supported the secession movement that carried Louisiana out of the Union and into a war they hoped would protect the culture and way of life they loved. Several family members played significant roles in the unfolding events, which changed forever their cherished lifestyles and left both Louisiana and the Confederacy in a state of defeat and ruin. During the long debate that preceded South Carolina's exiting the Union following Abraham

Lincoln's election to the presidency and before it was obvious to most that the Union would splinter, Mimi's husband and Aglae's son-in-law Richard Taylor expressed reservations about the wisdom of secession. As the crisis between North and South deepened and the fateful decision over Louisiana's relationship to the Union neared, Taylor's position shifted and he increasingly supported secession. As a prominent member of the state senate, he eventually became a leading proponent for Louisiana's immediate secession. Similarly, although many of their sugar-planter colleagues feared and opposed secession, Nanine's husband, Duncan Kenner, was also one of the state's leading proponents of secession. Kenner ran as a pro-secession delegate from Ascension Parish for a seat on the state's secession convention and was defeated. However, after Louisiana withdrew from the Union, he was chosen as one of six delegates to represent the state at the Montgomery Convention in Alabama where a provisional Confederate government was established under the leadership of Jefferson Davis.[234]

Though a majority of the Bringiers' sugar-growing neighbors held fears and reservations concerning the consequences of secession and a war with the North, most Louisianans and Southerners probably would not have disagreed with the Louisiana secessionists who proclaimed that the Yankees were "all a set of crying cowards!"[235] Yet, in spite of the worsening crisis and the uncertainty of the secession debate, most Louisianans attempted to live their lives as normal as possible. In early January 1861, just days before the state declared secession, Duncan Kenner was focused on his thoroughbred horses and the racing season in New Orleans as noted by the *Daily Delta*, which headlined "First day of winter racing, Mr. Duncan F. Kenner's *Sid Story* won a mile sweepstake . . . , purse $500."[236]

With Louisiana's secession on January 26, 1861, many individuals in the state continued their social activities much as they had done before secession with the exception of a change in theme. Most social events—dances, bazaars, and even weddings—were themed with military trappings and southern patriotism. Even after the outbreak of hostilities following the South's attack on Ft. Sumter, Louisiana's residents made efforts to carry on life with a degree normalcy. Although there were fewer extravagant affairs than before the crisis, smaller social events continued as before the war, with dinner parties remaining the most popular form of social intercourse among the planters. Noted British journalist William Russell cited one such gathering in his book *My Diary, North and South*. Russell visited the region in 1861 and wrote of his experiences, including a supper he enjoyed in early June at John Burnside's plantation with dinner guests consisting of prominent sugar planters including Duncan Kenner and M. S. Bringier.[237]

However, as the confrontation between the North and South worsened, the cherished lifestyle that Louisiana's planters had hoped to preserve began to change. With the crisis intensifying and the Confederacy's call for troops, the revelry and celebration that accompanied Louisiana's decision to secede gave way to the realization associated with the mustering of thousands to serve the cause. Although their social status offered them the opportunity to avoid participation, members of the Bringier family rallied to the Southern Cause and readily offered their services to the Confederacy. Among the first in the family to answer the call was Duncan Kenner. He represented the state at the meeting of seceded states in Montgomery, Alabama, which established the Confederacy. As a delegate to the meeting and later as a member of the Confederate government, Kenner was a loyal supporter of Jefferson Davis and took a strong nationalist approach, which favored a national government similar to that of the United States. Kenner went on to serve as a member of both the Confederacy's Provisional Congress and in its House of Representatives. As a member of the Confederate House, Kenner ascended to the position of chairman of the powerful Ways and Means Committee where he focused his attention on the Confederacy's finances. He supported protective tariffs and subsidies for internal improvements and pushed for the new nation to develop a diversified economy. Having spent much time in the North prior to the war, Kenner was cognizant of the power of the South's adversary and continuously supported efforts to strengthen the Confederacy, including conscription and all-out military preparation.[238]

With the capture of New Orleans in April 1862 by Union forces and much of south Louisiana's subsequent occupation by Federal troops, Kenner became convinced that only desperate steps by the Confederacy could stave off eventual defeat and Northern domination. Early in 1863, he approached his close friend (they shared lodgings in Richmond) and fellow Louisianian, Confederate Secretary of State Judah P. Benjamin, and later President Davis, with a daring plan to emancipate the South's slaves as a means of winning much needed diplomatic and military support from the Europeans. Many in the North and South believed that the slavery issue was the main obstacle stopping France and Britain from recognizing the Confederacy's independence and breaking the Northern blockade of the South's seaports. Although his dramatic plan was at first rejected by both Davis and Benjamin, as the South's military position grew increasingly desperate the Confederate leaders warmed to Kenner's proposal. Following Abraham Lincoln's reelection in late 1864, a series of crushing reverses on the battlefield, and with some leading Southern newspapers urging that slavery be sacrificed, if necessary, to save the Confederacy's national independence, Benjamin and Davis changed their opposition to Kenner's suggestion on emancipation and called on him to implement his plan. His plan required him to undertake a secret and dangerous mission to Europe and attempt to negotiate with the French and British for diplomatic recognition of the Confederacy in exchange for the South's abolition of slavery. Kenner was chosen not only because the plan was his and that he possessed the political skill and judgment to carry out the mission, but also because he had the respect of a large number of the Confederate Congress' members whose support was needed if the proposal had any hope of success. To assist him in winning the support of the Confederate diplomats already stationed in Europe, Kenner was given plenary power to negotiate with England and France for recognition. If successful in winning the support of the European powers, he was also given the authority to sell all the South's cotton to finance the purchase of arms for the Confederacy.[239]

By late 1864 when Kenner received authorization and provisioning for his secret mission to Europe, nearly all of the South's seaports were either in Union hands or under siege. Dedicated to his mission, he knew that the Confederacy desperately needed foreign assistance and that time was of the essence. To expedite his efforts, Kenner decided on a bold and dangerous journey in disguise through enemy territory to New York City where he avoided detection by Federal agents and secretly obtained passage on a German passenger liner for Europe. Upon completion of his five-week journey, Kenner reached Paris on February 24, 1865. He quickly met with the Confederate diplomats John Slidell and James Mason and informed them of his mission's goals. Although they strongly opposed the idea of emancipation, upon the realization that Kenner's authority superceded theirs, they promised their backing and assistance. Over the next couple of weeks, Slidell met with the French leader Napoleon III and Mason with British Prime Minister Lord Henry John Palmerston. Unfortunately for the hopes of the Confederacy, the meetings with the European leaders ended in failure. The French refused to act first and the British refused to risk a war with the United States because they were never convinced that Southern independence could be achieved beyond a doubt. Furthermore, the British Prime Minister noted that even if they wanted to assist the South at that time, the Confederacy's military options were exhausted and beyond revival. As Kenner and the two Southern diplomats worked to get the European powers to reconsider their decision, word reached them of Richmond's fall to Federal forces and that Gen. Robert E. Lee had surrendered on April 9 to Gen. Ulysses S. Grant at Appomattox Courthouse. All was lost! On June 20, 1865, Kenner visited the American Legation in Paris, took the oath of loyalty to the United States, and applied for an executive pardon. In late summer of the same year, Kenner sailed for home.[240]

Although Kenner's position in the Confederate government earned him considerable attention, the family member obtaining the greatest notoriety for his service to the Southern Cause was Richard Taylor. The son of former United States President and Mexican War hero Zachary Taylor, Richard (Dick) was a native of Kentucky who established himself on Fashion Plantation in St. Charles Parish. In February 1851, he married Myrthe Louise (Mimi) Bringier. During the long political debate leading up to secession, Taylor's moderate political views gradually shifted to where in 1861 he played a prominent role in promoting secession, including introducing the bill in the Louisiana legislature to call a convention to consider secession.[241]

When hostilities broke out between the North and South, Taylor offered his services to the Confederacy. Taylor had little experience or training in military matters. Entering the service as a colonel of a Louisiana regiment, there is little doubt that his political and family connections, including the fact that he was the son of a former president and the former brother-in-law of Jefferson Davis, helped him obtain his high rank.[242] A man of frail statue and "dandified mannerisms" who had been educated at some of the world's leading educational institutions including Harvard, Yale, and Scotland's Edinburgh University, Taylor quickly demonstrated to his superiors and men that he was a leader of outstanding ability and a brave and talented combat commander. By the end of the fall of 1861, he was promoted to the rank of brigadier general. Among his early assignments in the war was leading a brigade under Gen. "Stonewall" Jackson in the Shenandoah Valley and the Seven Days' campaigns in Virginia in 1862. Shortly thereafter, he was promoted to the rank of major general and assigned command of the Western District of Louisiana. While the leader of Confederate forces in western Louisiana, Taylor fought a series of battles against a large Union force under the command of Maj. Gen. Nathaniel Banks. Though his resources were inferior to Banks, in 1864 he defeated Banks at the battle of Mansfield in northwestern Louisiana, blunting the Union's efforts to take control of the state's vital Red River region. However, when his request for reinforcements after the battle went unheeded, which in his opinion prevented him from capturing the Union fleet on the Red River and, as he stated in his postwar memoirs, stopped him from recapturing "possession of the Mississippi, from the Ohio to the sea, and [undoing] all the work of the Federals since the winter of 1861," Taylor verbally assailed his superior Gen. Kirby Smith as "stupid" and threatened to resign from the service.[243] In spite of his pubic criticism of his superior officer, Taylor was promoted to the rank of lieutenant general and was given command of the Department of Alabama, Mississippi, and East Louisiana. As he had done throughout the war, he employed imagination and audacity to delay the Union capture of Mobile until the final days of the war. With the surrender of generals Robert E. Lee and Joseph E. Johnston, Taylor surrendered the last major Confederate army in the field east of the Mississippi River at Citronelle, Alabama, on May 8, 1865.[244]

Although Kenner and Taylor were the most famous members of the extended Bringier family to take part in the war, they were by no means the only ones to do so. Like so many families across both the North and South, many young males joined in the conflict, including Allen Thomas. A native of Maryland and a practicing lawyer, he was the husband of Aglae and Doradou's daughter Anne Octavie. Shortly after their marriage in 1857, the couple moved from Maryland to Louisiana to acquire a plantation. At the outbreak of the war, he worked to organize an infantry battalion in St. Landry Parish where he resided. When his unit was enlarged to regimental strength, he was elected colonel. His unit fought in various engagements associated with the Vicksburg campaign where he was captured when the rebel fortress surrendered on July 4, 1863. Quickly paroled by the Union, Thomas carried to President Davis the report from Gen. John C. Pemberton on his surrender of the Vicksburg garrison to General Grant. He was next assigned to the Trans-Mississippi Department where he was promoted at age thirty-four to the rank of brigadier general, making him one of the South's youngest generals. Six-foot tall and broad in his proportions, Thomas was known for his handsomeness and his personal bravery, often taking the lead in front of his men when ordering sorties and charges in combat situations. In February 1865, his skills and bravery earned him the command of an infantry division, which he led until the unit surrendered at the end of the war.²⁴⁵

In addition to experiencing the heartache of seeing her daughters' husbands leave their homes and families to serve the Confederacy, Aglae Bringier also suffered a mother's travail of seeing her sons and grandsons leave home to serve in the Confederate army. Three of the young male family members, including Aglae's nineteen-year-old son Doradou (familiarly Dadou), his older brother Louis Amedee, and grandsons Julien Bringier Trist and Nicholas Trist enlisted in Scott's First Louisiana Cavalry. Dadou and Nicholas enlisted together in early 1862. As cavalrymen they were expected to provide their own mounts. They were ordered to assemble in northern Mississippi at the town of Corinth and to have their mounts waiting for them when they arrived, so they shipped their horses ahead of themselves. When they reached their assembly point, they found their horses missing. Unable to find their mounts and unable to serve in a cavalry unit without them, they located Nicholas' brother Bringier Trist and joined his infantry unit, the Crescent Regiment. Within days of their joining the new military unit, it was involved in the battle of Shiloh on April 6 and 7, 1862. In the fight, Bringier was severely wounded in the arm and his brother and uncle carried him from the battlefield. After the battle, Dadou and Nicholas, obtained mounts and joined another mounted unit, Leed's Light Horse, which served as the body guard for Gen. Leonidas Polk. Eventually promoted to the rank of lieutenant, Dadou served as aide-de-camp to his brother-in-law Gen. Richard Taylor and later as an assistant to his older brother Amedee, who had been promoted to the rank of colonel. During the last year of the war, Dadou again served as aide-de-camp to a brother-in-law; however, this time it was for Gen. Allen Thomas and not Richard Taylor.²⁴⁶

Dadou's older brother Amedee's enlistment predated that of his brother. A fervent advocate of the Southern Cause, Amedee began his service to the Cause before Louisiana took the formal stance of secession. In the months leading up to Louisiana's leaving the Union, Amedee offered to assist Gov. Thomas Overton Moore in his efforts to prepare the state for the pending secession crisis. Moore appointed Amedee as a colonel in the state militia to facilitate the organizing of troops in western Louisiana. Having completed his recruiting services for the governor during the early days of the war and with his commission at an end, Amedee volunteered his services to the Confederate army as a private. However, the unit he joined was made up of volunteers and, as was the common practice of the time, the members of the detachment elected its own officers. When they voted for their leadership, Amedee was chosen as his battalion's major.[247]

His unit was assigned to the defense of New Orleans. When Farragut's fleet successfully passed the forts on the Mississippi, Amedee's battalion was among the last Confederate troops to evacuate the city. Before retreating from New Orleans, the unit was instrumental in transferring supplies to the new Confederate base at Camp Moore in Tangipahoa Parish about seventy miles north of the city. Shortly thereafter, when his unit was merged into the newly organized 28th Louisiana Volunteer Infantry Regiment, he was promoted to the rank of colonel. Not long afterwards when the regiment moved into Mississippi as part of the Vicksburg campaign, Amedee transferred to Company A of Scott's Cavalry, a unit that several family members had previously joined, including his brothers-in-law George and James Tureaud and a nephew, Emile Tureaud.[248]

Scott's Cavalry saw considerable action during the war, participating in major campaigns in Mississippi, Tennessee, and Kentucky. Usually serving as scouts for larger Confederate units, Scott's cavalry on occasion met in hand-to-hand combat with enemy units, including one fight when Amedee nearly missed being severely wounded or killed. When his unit engaged a Federal cavalry troop, the fighting became so close and intense Amedee and his men were forced to use their sabers. As he rose in his saddle to strike a blow with his saber, a Yankee foe fired at Amedee. The shot was so close that the ball passed between his legs and ripped into his saddle! As members of Scott's cavalry unit, Amedee and his fellow family members saw action in the major battles of Shiloh in April 1862, Perryville, Kentucky, in October 1862, and Murfreesboro, Tennessee, in late December and early January 1862-63.[249]

Amedee's military skill led to his promotion and made him a sought-after commodity by his superiors. In April 1864 when his former commander John Scott was promoted to command of Confederate cavalry units in East Louisiana and South Mississippi, he immediately sought Amedee's service. Writing his friend shortly after he received his promotion, Scott made the promise that "I will endeavor to organize a regiment for you" and added the plea that ". . . you must come immediately for I feel that I can not [sic] do without you."[250] Later in that same year, Amedee was recalled to Louisiana on the request of his brother-in-law Richard Taylor, who in May of 1864 was promoted to the rank of lieutenant general. Appointed Lieutenant Colonel of the 4th Louisiana Cavalry, Amedee quickly rose to the rank of colonel on January 7, 1865, and was given command of the regiment. Over time, the ranks of the unit greatly thinned through the attrition of war and desertions. Amedee, because of his prewar success in recruiting troops in the western part of the state, was given the task of rebuilding the regiment anew. Nearly three years of war had depleted the area of potential recruits. The lack of quality recruits forced Amedee, like other rebel commanders in the region, to turn to the use of stragglers, enlistment-shirkers, the elderly, and the young to fill out the ranks of his unit. To get them ready for combat, Amedee imposed a rigid regimen of discipline, drilling, training, and discerning management. His efforts were so successful in raising the military efficiency of the unit it was renamed the 7th Louisiana Cavalry or Bringier's Cavalry. Because of the new unit's readiness standing, it was one of only four regiments placed under the command of Brig. Gen. Joseph Lancaster Brent.[251] General Brent in turn was a subordinate of Brig. Gen. Allen Thomas, Amedee's brother-in-law. The quality of Amedee's leadership and training was distinguished by the discipline and cohesion of his unit—no easy accomplishment at that time in the war. With its ten companies continuously maintaining full complements, the unit contrasted drastically with the disorganized, depleted, and exhausted nature most Confederate military units in the field exhibited during the closing months of the conflict.[252]

Known for his bravery and abilities on the battlefield, Amedee in most instances insisted on a high level of preparedness and order among his men. He was not opposed to subjecting them to stern discipline to sharpen their military skills. No one in his command was safe from his demanding standards. Even his younger brother Dadou, who served as one of Amedee's military aids, was not spared. A careful manager of his scarce supplies, Amedee took great offense when he discovered his younger brother copying a document over because the first copy contained a clerical error. At the time paper was an extremely rare commodity, with individuals often answering letters by writing on the back and margins of the original pages. Amedee's tongue lashing of his brother over the wasting of a sheet of paper was so severe, that the young lieutenant immediately resigned his position on the colonel's staff and transferred to another command. Colonel Bringier's demanding standards not only applied to his troops; he also set high expectations for the inhabitants of the areas that fell under his command. Notably, he demonstrated little patience in dealing with guerillas, jayhawkers, and deserters who frequently took part in indiscriminate plundering, killing, and arson in areas of the state with minimal military or law enforcement presence. When captured, Colonel Bringier's justice in dealing with these criminal elements was often merciless and frequently resulted in summary hangings.[253]

Amedee's demanding measures did not come without a cost and criticism. In the fall of 1864, while serving as the commander of the 4th Louisiana Cavalry, Amedee requested reassignment to a different unit because of a dispute with an unnamed superior. Following Amedee's execution of an individual he labeled as a "colored jayhawker," the superior officer took the unusual step of publicly criticizing his fellow Confederate officer for possibly acting hastily. When the criticisms appeared in the local newspaper the *Alexander Democrat* on October 27, 1864, Bringier asked to be relieved of his command and reassigned. On January 7, 1865, Amedee was promoted to colonel and given command of the 7th Louisiana Cavalry. Whether the promotion, which came two months after he wrote his letter, was the result of Bringier's request to be reassigned is not known, however, his transfer to a new unit did not resolve Amedee's issues with his Confederate colleagues. Whether it was his austere personality or the strict standards he imposed upon his troops, during the final months of the war Amedee suffered a loss of confidence in his ability to carry out his orders. The issue became so serious that on March 18, 1865—slightly more than two months after taking command of his unit—he submitted his resignation to his commanding officer Brig. Gen. Joseph L. Brent bitterly claiming that he had "lost with most of my officers that influence which a commander should have over his subordinates; and . . . I am surrounded by a set of ungrateful, contemptible, envious, ambitious, beings with whom I cannot condescend to serve."[254]

Although a brave and tenacious military leader, as demonstrated by his resignation remarks, Amedee was also a serious and humorless individual who at times brooded over not only the circumstances and people with whom he had to work but also the South's military reverses and its potential for success in the war. A loyal supporter of the Confederacy, Amedee nevertheless was a realist who harbored doubts about the South's chances for long-term success in its contest with the more industrialized North. Even in the early days of the conflict when the South was achieving some success on the battlefield and many Southerners were filled with enthusiasm and confidence, Amedee's communications with family members imparted a melancholy about the future. It is possible that the gloomy forebodings coming from such a faithful Confederate in some degree influenced Duncan Kenner's bold decision to press President Davis and Secretary Benjamin to consider emancipation as a means of winning European diplomatic recognition and military support.[255]

Amedee's despondency was influenced by the loss of close friends in the war—including his wife's brother George Mather Tureaud who died shortly before the battle of Shiloh—the deteriorating military standing of the Confederacy in general, and the gloomy situation at home in Ascension Parish which weighed heavily upon him. Since his father's death, Amedee, in a partnership arrangement with his mother, managed the day-to-day operations of the Hermitage estate. With the onset of the war, he placed the plantation under George Zenon Trudeau's care, the brother of Gen. James Trudeau. The estate remained safe during the early months of the war and on occasion Amedee was able to get leave from his military duties and returned home to visit his family. However, with the Federal capture of New Orleans, the situation changed quickly. Following the capture of the Crescent City, Union forces did not immediately occupy the lands along the river. A powerful Federal naval force did advance up the river past the Bringier plantations and captured Baton Rouge and Natchez. Fortunately for the planters who lived along the river, Federal troops did not seriously bother the plantations along the waterway. Federal naval forces on the river were content with patrolling the waters of the Mississippi River and did not attempt to seize the land along the river.[256]

During the early months of the Federal invasion of Louisiana, the worst scare that the Bringiers endured occurred when Farragut's fleet first passed the forts that were supposed to protect New Orleans from such an event. With the exception of Aglae who was away on a visit to the Hermitage, at the time the Union fleet succeeded in piercing the Confederate defenses, several Bringier women were staying in New Orleans, while many of the family's men were away serving the Confederacy. With the church bells announcing the eminent arrival of enemy forces, the Bringier women feared for their safety and immediately made plans to leave the city. No one knew what would happen when the enemy arrived in New Orleans. As many family members as possible gathered at Melpomene and prepared to evacuate. Getting out of the city was no easy task since all means and arteries of transportation were clogged with rebel troops hauling their supplies and equipment out of the city. Fortunately, Maj. (later general) Allen Thomas was able to get the party of approximately twenty, and their servants, passage on a steamboat that was specifically detailed by Confederate authorities to evacuate the families of rebel officials. With the city's riverfront bathed in dense black smoke from the burning of cotton and other valuable commodities so as to keep them from the hands of the approaching enemy, a feeling of deep gloom settled over the little party as their steamer sailed away from New Orleans. Not only were they leaving behind their beloved city and homes, in their hearts they realized that they were leaving behind a way of life that they had loved and enjoyed for as long as they could remember.[257]

In spite of being advised to move to safer areas, several of the group chose to disembark at the Hermitage plantation. Some, such as Nanine Kenner who lived nearby on her Ashland plantation, returned to their homes. Others in the party continued on until the vessel reached Red River landing in central Louisiana. There they established a small colony of sorts near Bel Cheney Springs for various family members.[258]

There was a period of near normality at the family's plantations along the river. Though the Union navy had brazenly moved its warships up the river past the plantations and its gunboats frequently patrolled along the river, they gave little notice to the homes and plantations located along the waterway. The situation became so normal that Nanine's husband Duncan Kenner, a high-ranking member of the Confederate congress, was able to return to Ashland plantation a few miles upriver from the Hermitage. There he was able to sit on his upper veranda and watch Federal gunboats steam up and down the Mississippi River. However, the situation radically changed in the summer of 1862. On the afternoon of July 27, Union troops disembarked from a river steamer and raided Kenner's Ashland plantation in an unsuccessful effort to capture the rebel congressman. At approximately the same time, the commander of Federal naval forces on the river, David Farragut, began a policy of aggressive response to Confederate sniping against his vessels as they traversed the river near Donaldsonville and the Hermitage.[259] In July, he sent a brief note to the residents of Donaldsonville in which he simply stated the stark warning that "Every time my boats are fired upon I will burn a portion of your town."[260] Unfortunately for the area's residents, Farragut proved to be a man of his word. When the sniping continued, Farragut issued a final warning and on Saturday, August 9, 1862, three Union gunboats bombarded the town for about one hour. The bombardment and a follow-up raid by Federal troops, with orders to torch whatever was left that could comfort the rebels, left the town two-thirds destroyed. In spite of the destruction, sniping continued to plague Union ships in the Donaldsonville area. In response, Federal commanders ordered an increased military presence and response in the region, including increased attention being paid to the area's large plantations. The Hermitage became subject to raids and other harassing efforts by Union troops. From time to time Federal cavalry units quartered on the estate and, whenever they felt it necessary, forcibly requisitioned supplies, fodder, wood, and livestock.[261]

The arrival of Federal forces and the threat of increased military action in the river parishes made life even more difficult for the Bringiers and other locals. In the days shortly before Farragut carried out his threat to burn Donaldsonville, Stella Bringier clearly felt the distress of the Union threats of retaliation. She wrote with great despair to her husband Amedee who was away serving in the Confederate army that:

> We cannot keep back the guerillas, they are determined fellows and doing what they have been ordered to do . . . but what are we to do? . . . I know not. I am sick with torturing my brain to try and find out what would be the right course to pursue. If you were only here to tell me. With four little children where can I go to, and how? Gunboats have been passing down at a rapid rate since yesterday. Something is certainly expected to take place.[262]

Stella's prediction of pending action was no doubt made before she and her mother-in-law Aglae had received word of the raid on Ashland the night before she wrote her plea to her husband. The Federal's raid on the Kenner plantation, the arrest of all the white men on the property, and the confiscation of all the movable valuables on the plantation caused the family great alarm. To the white Southerners of the period there were few fears greater than the thought of having white women and children left alone on an estate with hundreds of unguarded slaves. When Aglae learned of the Federal attempt to arrest her son-in-law and that her daughter and grandchildren were left at Ashland without what she considered to be adequate protection, she immediately went to her daughter's aid. Arriving the day after Union troops had left the Kenner property, Aglae found that the house had not been burned as she had heard and that Nanine and her children were safe and protected by several loyal slaves. Aglae, Nanine, and the Kenner children returned to the Hermitage for a short time. Fearing that the events at Ashland could be repeated at the Hermitage and wanting to help the Kenners reach safety in an area where they would be secure and able to reunite with Duncan,

Mrs. Bringier packed her things, joined the small caravan of the Kenners and their house servants, crossed the river on the Donaldsonville ferry, and headed west into the Lafourche country, behind Confederate lines. Reuniting with Duncan, the family traveled to Houma where they stayed with Duncan's cousin, William Minor, Jr., at his large and newly constructed Southdown Plantation. After three months, Union forces threatened the Houma area and the family moved again. This time they traveled through the Teche country to near Opelousas where Aglae's daughter Octavie and her husband Allen Thomas lived on their New Dalton plantation. Feeling safe from Federal forces and comfortable because of the small exile colony of planters from the river parishes, which had settled in the area, they were soon joined by other family members including Myrthe Taylor and Benjamin and Marie Elizabeth Tureaud. Exiled from their Tezcuco estate, the Tureauds joined with the Kenners and together rented a large house and estate at Moundville, not far from Washington, Louisiana.[263] Eventually the Kenners moved again farther north to Natchitoches on the Cane River deep behind Confederate lines where living conditions were pleasant and less refugee-like.[264]

With the military situation at Moundville growing less secure, Aglae and her daughters Myrthe Taylor and Octavie Thomas moved to Alexandria in October 1862. She continued to reside with her daughters until Octavie made the perilous journey to visit her husband who was serving with the Confederate forces near Vicksburg. Gen. Ulysses S. Grant had maneuvered his forces ever closer to the rebel stronghold on the Mississippi River and the pregnant Octavie wanted to visit with her husband Thomas before it became too late for her to be able to get through the military lines to see him. With Octavie's departure, Aglae left Alexandria and moved further north to stay with the Kenners in Natchitoches.[265]

While her mother was making plans to move to be with Nanine and her family at Natchitoches, Octavie set out on her trip to see her husband. She had made the arduous trip on two previous occasions. However, with Federal forces increasing their pressure around the rebel fortress on the Mississippi River, her third trip in May 1863 was much more dangerous. Having previously sent the thirty-two slaves from her New Dalton estate to Nacogdoches, Texas, where they were to be hired out under the supervision of a trusted overseer, she traveled with only her young son Allie, his nurse Olivia, and an unnamed companion to the banks of the Mississippi River. Once there, she and her small party made their way cautiously across the swiftly moving and dangerous river in a small skiff to Natchez. Fortunately, during her trip she met a young Confederate officer who was attempting to return to his post at Vicksburg who assisted her during the most dangerous portion of her trip. From Natchez the small party moved on to Brookhaven where her husband, wearing civilian clothes, met her and took her through the Confederate lines to Vicksburg.[266]

Octavie's arrival at the Confederate fortress on the Mississippi did not mark the end of difficult times for her. Arriving in Vicksburg in May just as Federal forces under the command of General Grant escalated their campaign against the rebel fortress, Octavie found the town in a wretched condition from the ravishes of the Yankee siege. As Grant tightened his grip on Vicksburg, Octavie and her husband became increasingly concerned for the family's safety. Federal artillery increasingly shelled Vicksburg. The bombardment required Octavie and her son to take refuge among the many small caves the town's residents dug into the hillsides for protection from the falling bombs.[267]

During the first days of June 1863, Octvaie and her small party sadly said goodbye to her husband and moved to a plantation owned by friends on the Yazoo River about fifteen miles from Vicksburg. While at the plantation she again faced danger when Union troops raided it. While in the process of carrying off the plantation's valuables, the Federal unit was attacked and driven off by a detachment of Confederate cavalry. Ironically, the rebel force was under the command of another member of the Bringier family, Col. Robert C. Wood. The Confederate colonel was the husband of a niece Wilhelmine Trist and the future father of the Bringier family's genealogist Trist Wood. Realizing that the fighting in the area would only intensify, Colonel Wood advised Octavie to leave the area immediately. He procured mules and an ambulance and detailed a young officer named Bowman to escort her and her small entourage to safety. She made her way to Canton, Mississippi, where she was again advised by Confederate officials that the area was not safe and that she needed to move to safer ground. Heading south, she eventually made her way to Mobile, Alabama, where upon her arrival she learned of the fall of Vicksburg. During her move to escape the fighting around Vicksburg, Octavie lost all contact with her husband and family members. Reaching south Alabama, she was completely in the dark over the fate of nearly everyone and everything she held dear. She moved from Mobile across the bay to Point Clear where the Confederates maintained a hospital for troops wounded in the Vicksburg campaign. No doubt she hoped that her presence at the hospital would at least present the possibility of her finding her husband if by chance he had been wounded in the campaign. Travel while pregnant under such adverse conditions and the worry and stress associated with being alone and out of contact with her family no doubt weakened her health and that of the baby to such a degree that shortly after being born on August 12, 1863, her son tragically died. Although she probably remained in Alabama for sometime after the death of her baby, it is not known when she returned home to Louisiana.[268]

Although large numbers of planter families who lived along the Mississippi River and Bayou Lafourche joined Aglae and Nanine in fleeing their homes upon the approach of Federal forces, others in the region attempted to sit out the situation and waited to see what life would be like once Union troops arrived in the area. Even after the arrival of Federal troops in the vicinity, M. S. Bringier decided to remain at his Houmas/Brulé properties and did so for the remainder of the war. In spite of the fact that her mother and sister evacuated the area and that she feared for her own and her children's safety, Stella Bringier was determined not to leave the Hermitage. She remained with her children at the family's home and attempted to keep the estate in operation. Not unexpectedly, the area's occupation by Federal troops and having them frequent the plantation caused tension for Stella and her four children. As military activity along the river in the Donaldsonville area increased, the family's situation at the Hermitage worsened appreciably. Tensions escalated until an incident occurred which directly threatened the life of Stella and her children and forced her to make the difficult decision to abandon the plantation. Apparently, as part of Farragut's operation to retaliate against sniping by Confederates along the banks of the Mississippi, a Federal gunboat passing in the river opened fire with one of its cannons upon the Bringier house. Though the shot missed the structure, it cut a large limb from a great live oak located near the building, which crashed into the house crushing a corner of the building. Fortunately, no one was injured by the cannon shot. Stella now realized the extent of the growing danger to her children and decided that she had to get them away from the plantation to avoid additional danger. She and her children left the Hermitage on September 5, 1862.[269]

As had other members of the family, Stella and the children moved to the Louisiana interior behind Confederate lines. After a little over a year with Federal forces gradually strengthening their hold on the area around the Hermitage, Stella felt that the situation at the plantation had become safe enough for her to return home. However, on

Conflict and Loss

November 20, 1863, the very night she returned to the Hermitage, the situation changed dramatically. After midnight, a detachment of the 1st Louisiana Regiment of the Union army under the command of 2nd Lt. George A. Mayne moved onto the estate and seized it along with anything of value that could be found. [270]

Before he left for military duty, Stella's husband Amedee arranged for George Zenon Trudeau to manage the operations of the Hermitage. A longtime family acquaintance before the war, Trudeau had spent time at the estate performing accounting work for Aglae. In the period leading up to the raid of November 20, 1863, Trudeau and Stella Bringier had more or less successfully handled the enemy soldiers' presence in the area. Though the proximity of Federal troops led many of the slaves in the region to take advantage of the opportunity to flee their plantations for the Union lines, the slaves at the Bringier plantations largely remained loyal and did not runaway when the opportunity arose with the arrival of Federal troops. Even in July 1862 when Union soldiers raided area plantations, including Ashland in an attempt to arrest Duncan Kenner, the family's slaves did not abscond. Following the raids, Stella wrote her husband and noted with some relief that "not one of [their] negroes" had left the plantation. Similar loyalty was demonstrated at Ashland where a much larger number of slaves resided. After the war Kenner's daughter Rosella recounted the terrifying events of the Union's attempt to arrest her father and noted that "Some of the negroes went with the Federals when they left, but the majority remained at home." She added that because the Federal troops mistreated the bondsmen at Ashland "the negroes were not tempted to join them & encounter the fortunes of war." Even as late as 1864, Kenner commented that "I am glad to learn that so many of my Negroes, if I can still call them mine, are at home."[271]

Nonetheless, the loyalty of their slaves was not sufficient to keep the workers on the plantations forever. Like most of the region's plantations, after the first Union raids in the area, the Hermitage and the other Bringier plantations along the river were occasionally visited by Federal troops who requisitioned whatever supplies they needed, including fodder, wood, and livestock. During their visits they generally respected the family's home and belongings—that is until the raid of the night of November 20, 1863. Arriving by steamboat in the early hours between midnight and daybreak, they seized the family's personal property. Zenon Trudeau strongly protested to the officer in charge. Explaining that he was simply "acting under orders," the officer ignored

Trudeau's objections, placed him under arrest, and held him in a slave cabin. Everything of value on the estate that could be moved, including the estate's stock of perique and snuff tobacco valued at an estimated $180,000, all the slaves—men, women, and children—farm animals, wagons, carts, etc. were taken on board a steamboat and removed to Donaldsonville. According to family recollections, the estate's slaves, including the house servants, were "corralled and driven off without even permission to say good-bye to their mistress." With the plantation stripped of its workers and furnishings, Stella once again realized that the place was not safe for her family. Within a day or two of the raid, Stella and her children abandoned the Hermitage for the second time. Neither she nor any other member of the Bringier family would enter the home again until after the war.[272]

The Hermitage did not remain vacant for long. In July 1862, Congress passed a law allowing for the seizure of the property of individuals disloyal to the United States. Though the act was enforced in Louisiana in those areas comfortably under Union control such as New Orleans, in the contested area around Donaldsonville enforcement of the law was sporadic. However, with it being situated on the Mississippi River just across from Donaldsonville and because of the Bringiers' notoriety, the Hermitage was a tempting prize for the Federal authorities. Within days of Stella leaving the Hermitage, Capt. C. W. Cozzens of the Federal Bureau of Abandoned Property took possession of the plantation. Federal officials wanted to return the many abandoned plantations in the area to production. Anxious to see the return of the region's sugar estates to productivity so as to provide the Union with revenues, food, and work sites for the many former slaves who had left their home plantations and were roaming the area, Federal officials offered the properties under their control for lease. Expecting to be able to duplicate the financial success enjoyed by southern planters in the years prior to the war, many northerners hastened to the area to take advantage of the economic opportunity offered by leasing an abandoned estate.[273]

Confiscated estates in southeast Louisiana's sugar region were placed under the supervision of Capt. C. W. Cozzens. Although each property was placed under the control of a manager who was expected to return the plantation to productivity, Cozzens was often personally involved in the supervision of the properties. According to the recollections of Bringier family members, the Union officer's involvement with the Hermitage was much more than simply that of a supervisor. After the war, family members claimed that it was Captain Cozzens

Conflict and Loss

who personally removed many of the family's personal household articles and family treasures, including their much loved cut glass and rare book collections. For years after the war, family members complained that they knew from reliable sources that some of the rare books from the collections at the Hermitage remained objects of admiration for visitors to Cozzens' home in New Orleans.[274]

Eventually, Federal officials leased the Hermitage plantation to James Baxter. Like many of his counterparts, Baxter looked upon his stay in the South as a purely moneymaking opportunity. Leasing several estates, including the "Butler Place" on Bayou Black, Live Oak or the "Quitman Place" on Grand Bayou, Riverton plantation, and the Hermitage, Baxter concentrated on maximizing profits for himself and his financial backers who remained in the North. Other than generalized comments in Bringier family materials that Baxter was a northerner, surviving records do not reveal many particulars about him. Even the identity of his home state is unknown. Originally, he carried on business from his Riverton plantation in Ascension Parish. However, once he obtained the lease of the Hermitage, he moved his operations to the Bringier property and used it as his home and business headquarters. As was the case with many of his colleagues from the North who leased confiscated sugar plantations and expected quick profits, Baxter discovered that the management and profitable operation of a sugar plantation was not a simple task. Most notable of the difficulties faced by the lessee planters was the management of their labor forces. The arrival of Union forces in the region resulted in thousands of former slaves abandoning their plantations either as runaways or, as in the case of the Hermitage's bondsmen, being removed by the military as contraband of war. Sugar production required a large workforce. Federal officials, including both of the region's Union commanders during the war, Gen. Benjamin Butler and his replacement Gen. Nathaniel Banks, attempted to develop workable labor systems which freed the workers from slavery but required them to work the plantations and bring in the crops. Planters were required to pay their workers with either food and shelter or wages. Few were satisfied with this early venture into wage labor. Neither the workers nor planters had a clear understanding of the new labor relationship. Sugar production during Federal management suffered and never came close to reaching prewar production levels.[275]

Additionally, sugar planters had to overcome a serious shortage of the draft animals needed to produce a crop. As in the case of the forays on the family's Hermitage and Ashland estates, raiding troops often seized the animals found on the properties for their own use. Similarly, plantation support buildings and structures, including the vital levees, barns, and sugarhouses were all damaged or destroyed by neglect or wartime actions. These difficulties were made worse by wartime inflation, which beset the area and made prices of common necessities and farm equipment rise beyond the reach of many planters. The problems faced by sugar growers during the war became so overwhelming that many abandoned, either partially or completely, crop growing. In an attempt to work around the problems associated with producing a sugar crop, many planters diversified their production in order to grow other food items such as sweet potatoes, corn, peas, Irish potatoes, and other leafy vegetables to meet their own food needs as well as to meet the wartime food needs of the area. At the Hermitage, Baxter also diversified his efforts. In addition to growing corn, he stocked the estate with hogs to cut down on his food costs and set up a sawmill to harvest the estate's ancient and slow-growing cypress trees for lumber. Although he increased the variety of crops he grew on the property, Baxter never completely abandoned sugar production at the Hermitage. Meeting, however, with only limited success in its production and processing, he was forced to reduce greatly the acreage allocated to the crop. In May 1865, he reported to a financial backer that he had given up on his sugar-refining efforts for the year and would instead "save every stick of cane this year for seed."[276]

Baxter's most ambitious diversification was with cotton. Wartime shortages had driven the price up dramatically. Despite the unsuitability of south Louisiana's climate for the crop, their lack of success with sugar and the high price demanded by cotton coupled with the urging of Federal officials to expand the cotton acreage in the area led Baxter and his colleagues to focus on its cultivation. Though it is not certain from his papers if Baxter was referring to the product of all of his leased estates or just the Hermitage, in a letter to his backers in early June of 1864, he boasted of having 1,000 acres of cotton and 600 acres of corn under cultivation and had great expectations of sizeable profits.[277] It is probable that these optimistic projections for profits and acreage were somewhat exaggerated for in the same letter he had to explain to his financial backer the huge loss he had sustained as a participant in General Banks' ill-fated Red River campaign. In addition to defeating Confederate forces Bank's 40,000 man invasion of central Louisiana had a secondary goal, to earn great wealth for its participants and civilian backers who hoped to profit from seizing cotton which reportedly filled the Red River region.

The expectations for the invasion were so optimistic that the Treasury Department issued special permits to limit the number of civilians who hoped to follow the army in search of cotton profits. James Baxter considered himself one of the "lucky ones" for having obtained a permit, which allowed him to participate in the expedition. Acquiring 125 bales of cotton near Bank's base at Alexandria, Baxter, as everyone else who participated in the risky venture, including Martin Gordon, Jr., lost his entire investment when he could not obtain quick transport for the material to New Orleans and saw it burned during the Federal retreat down the river after Richard Taylor defeated Banks.[278]

While most lessee planters faced security difficulties, Baxter managed to avoid the issue while he worked the Hermitage. Although the area along the Mississippi River was largely under Union control during most of the time that the Federal government leased the large plantations along the river, there was a constant threat of attack by Confederate marauders. The rebel raiders often singled out those northern entrepreneurs who leased the property of loyal Confederates for severe punishment. As late as March 1864, the New Orleans newspapers noted with alarm the frequency of raids in the area along the river between New Orleans and Donaldsonville. Fortunately for Baxter, the proximity of the Hermitage to the Federal base and fortress of Fort Butler at Donaldsonville placed the estate well within the protective range, as Stella Bringier discovered, of the Yankee gunboats' deadly naval guns that often anchored nearby. Situated on the east bank of the river nearly opposite Donaldsonville, the plantation was a convenient and strategically significant site to base troops. Nearly the entire time that Union forces occupied Donaldsonville, sizeable numbers of Federal troops remained based at the Hermitage. Among the units spending extended periods of time at the plantation were the 31st Massachusetts Infantry and the 3rd Rhode Island Cavalry. These units were responsible for protecting east bank properties from the Donaldsonville area south to Bonnet Carré from raids by Confederate guerilla units that frequented the area from their bases north and west of lakes Maurepas and Pontchartrain. It was from the Hermitage that Federal forces launched patrols in order to block the attacks in the river region from rebel raiders. Union troops continued these blocking operations until the end of the war. The soldiers' continued presence at the plantation allowed Baxter to occupy the property in relative safety until the end of the war.[279]

The only known image of Marius Pons Bringier's White Hall plantation was depicted in an early painting by his son-in-law Christophe Colomb (who included himself in the image toward the bottom left of the painting). Courtesy of The Historic New Orleans Collection, accession no. 1971.40.

Master of Hermitage Plantation, Michel Doradou Bringier made the estate the centerpiece of his extensive land holdings along the Mississippi River. Courtesy of The Historic New Orleans Collection, accession no. 1970.11.35.

At right, Aglae Bringier's uncle, Bishop Louis William DuBourg, was an influential member of both the family and the Catholic Church in America. Courtesy of the Archdiocese of New Orleans.

At left, Aglae Bringier, wife of Doradou Bringier, in her elder years. Upon his death in 1847, she assumed control of the family's holdings and spent much of her time at the family's spacious Melpomene house in New Orleans. Courtesy of The Historic New Orleans Collection, Trist Family Papers.

At right, Doradou and Aglae's eldest child, Marius St. Colomb (M. S.) Bringier operated the family's Houmas and Brulé properties. A talented inventor, he obtained eleven different patents on devices designed to improve the production of sugar. Courtesy of The Historic New Orleans Collection, accession no. 1970.11.11

At left, Duncan F. Kenner, who married Nanine, daughter of Doradou and Aglae Bringer. The couple lived at Ashland Plantation and eventually acquired ownership of most of the Bringier family's plantations. Portrait by Rudolf Bohunek. Courtesy of the Louisiana State Museum.

At right, a former slave of the Bringiers on their Houmas plantation, Pierre Caliste Landry operated the plantation store on the estate and after the Civil War was elected mayor of Donaldsonville, Louisiana, making him among the first African Americans to hold such a position in America. Courtesy of the Amistad Research Center.

At left, son of former U.S. President Zachary Taylor, Confederate hero Richard Taylor married Doradou and Aglae's daughter Myrthe. The couple lived at their Fashion plantation in St. Charles Parish. Courtesy of the Library of Congress.

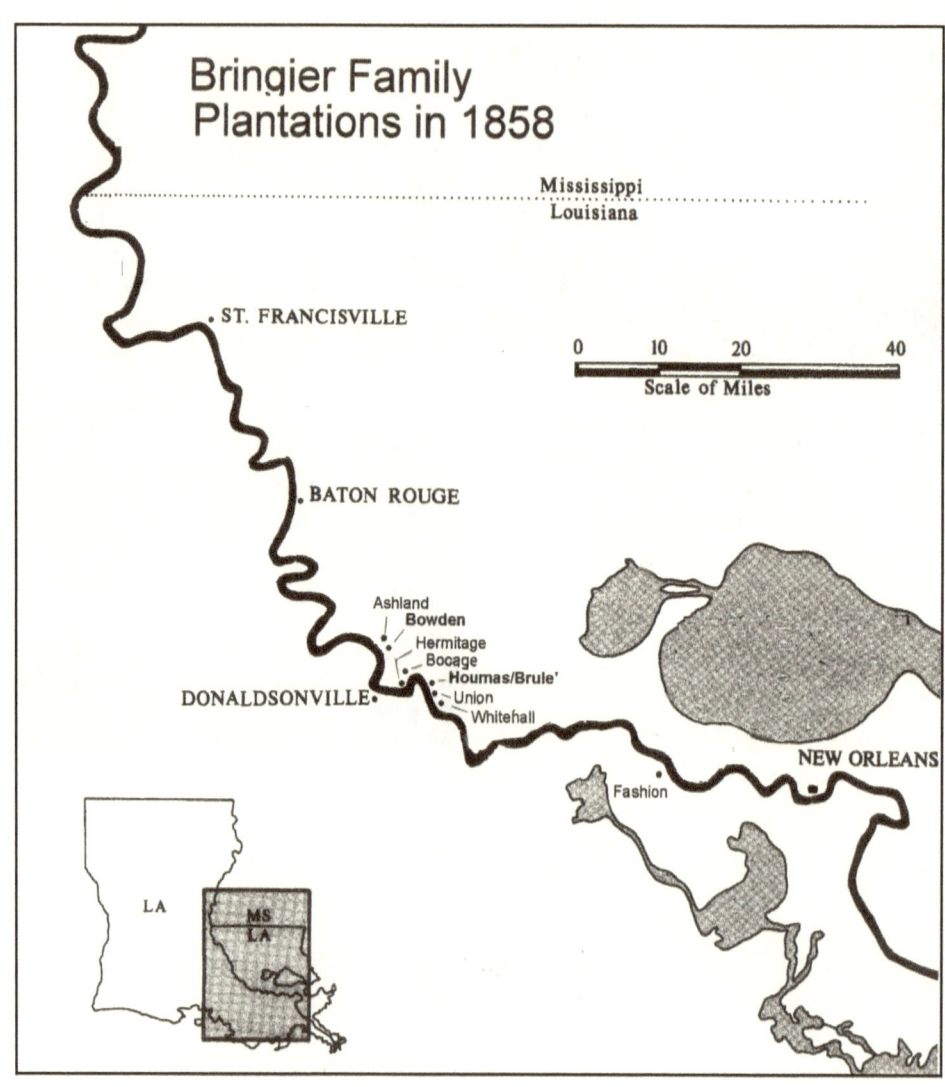

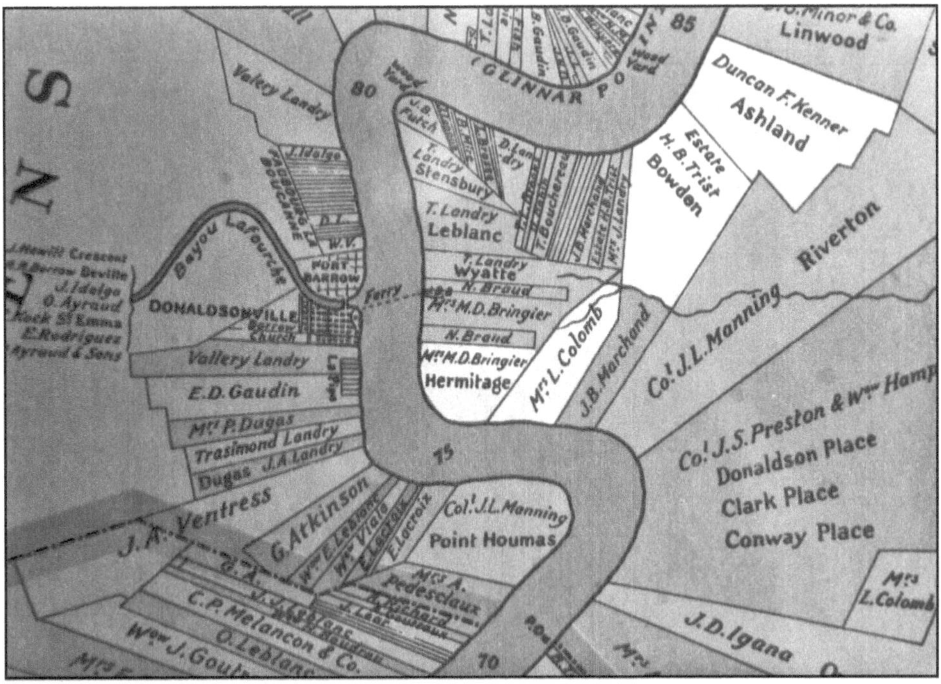

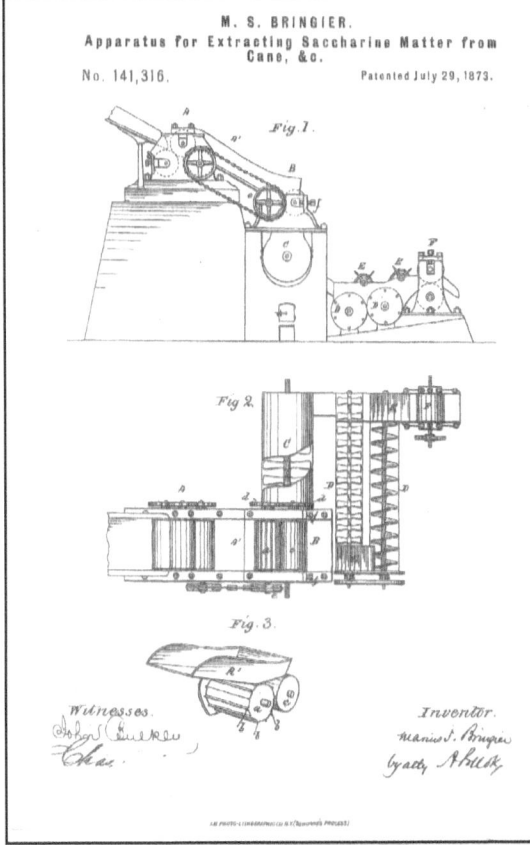

Above, section from Adrian Persac's "Norman's Chart of the Lower Mississippi" showing the Hermitage, Ashland, Bowden, and Mrs. L. Colomb's Bocage plantations, and their locations near Donaldsonville. Courtesy of the Library of Congress.

At left, patent design for M. S. Bringier's "Apparatus for Extracting Saccharine Matter from Cane." This 1873 patent was among the most elaborate of his designs.

Opposite page: top, the largest and grandest of the Bringier family's plantation houses, Ashland was the home of Duncan Farrar Kenner and his wife Nanine Bringier (photo by the author); middle, constructed in 1855 as the home of Benjamin Tureaud and Marie Elizabeth Aglae Bringier, Tezcuco was destroyed by fire in May 2002 (courtesy of the Louisiana State Library); bottom, Melpomene, the Bringiers' city home, which was demolished towards the end of the nineteenth century (courtesy of the Louisiana State Library).

This page: above, Hermitage plantation house showing the cistern base and rear addition added by Doradou Bringier. The addition was removed in the early twentieth century (courtesy of the Louisiana State Museum); below, the Hermitage's sugarhouse just after the Civil War, note the main manor house at the far right (from Thomas West Smith's *The Story of a Cavalry Regiment* . . . [1897]).

Located in the Catholic cemetery in Donaldsonville, the once grand tomb constructed by Doradou Bringier holds the remains of many of the members of the Bringier family and today shows the fatigue of age. Photo by the author.

Chapter 9

The End of a Way of Life

The end of the war brought a final realization to the Bringiers that the way of life they had cherished and defended for their entire lives was gone forever. During the war, members of the close-knit family scattered across the state and the South—some as soldiers and Confederate officials and others as refugees. With the war over, the natural want of all was to return to their homes and to rebuild their shattered lives. The devastation suffered by Louisiana and the South during the conflict made rebuilding a difficult task that many once prosperous planters found impossible to achieve. Although much of Louisiana had been spared large-scale military action during the war, the state paid a high price for its membership in the Confederacy. Only Virginia, South Carolina, and Georgia endured greater destruction and casualties during the war than Louisiana. Not only did the state lose thousands of its brightest and bravest sons from wounds and sickness during the conflict, it suffered devastating financial damage from emancipation which resulted in a loss of approximately one third of its economic wealth. Other costs to Louisiana included one half of its farm livestock and the destruction of two thirds of its farm equipment and machinery necessary to produce the state's sugar crop. Besides the losses on the plantations, the state's banking and railroad industries were also near ruin. These factors were so damaging that by the end of the war, the number of sugar-producing plantations dropped from a prewar total of 1,200 to approximately 175. According to historian John Winters in his study of the Civil War in Louisiana, when all financial losses from the conflict were counted up, Louisiana lost more than half of its prewar wealth.[280]

For months leading up to the event, the Bringiers and most other white Southerners could portend the inevitable defeat of the Confederacy. When the end actually came in the late spring and early summer of 1865, supporters of the Cause were numbed and dismayed by the news. Relieved that the killing, dying, and destruction of the previous four years were at last over and that their surviving loved ones would soon return, the Bringiers recognized that much work needed to be done before their lives could even begin to return to a semblance of normality. They knew that the institution of slavery, on which so much of their prewar lifestyles depended, was gone forever. They also knew that their losses included much more than their slaves and the personal and farm items hauled away during the Union raids on the family's properties. Their local center of business, commerce, and social activities, Donaldsonville was in ruins as well, as were many of the region's bridges, roads, public buildings, and the all important levee system. Perhaps most troublesome of all was the fact that the family's home place, the Hermitage, remained in the possession of the Federal government and continued to be occupied by the Unionist lessee James Baxter.[281]

Additionally, there was some uncertainty over what would be the Federal government's policy toward the former Confederates. In May 1865, Pres. Andrew Johnson issued proclamations setting forth his reconstruction plans. These included the restoration of all property rights, excluding the return of slaves, to those former rebels who took oaths of allegiance to the Union and accepted emancipation. However, Johnson's plan also required Confederate officials and owners of property valued in excess of $20,000 to apply individually for presidential pardons. The requirement of individual pardons led many in the North and South to assume that the president intended to carry out his long-noted aim of ending the political and economic power of the planter elites of the defunct Confederacy. During the war, Johnson publicly discussed the need to "punish and impoverish" the planter class in the South for leading the rebellion and proposed the possibility of confiscating the South's large plantations to divide up for sale to "honest farmers." Such comments by the new president only added uncertainty to the lives of the Bringiers and their planter colleagues.[282]

By the time Amedee Bringier, Richard Taylor, Allen Thomas, and the various other family members who served the Confederacy made their way home after their units disbanded, some of the great uncertainty and anxiety of what the postwar period had in store for them were diminishing. By the end of summer of 1865, many white Southerners were increasingly encouraged by President Johnson's actions in dealing with the vanquished rebels. In spite of opposition in the North, Johnson reversed his earlier plans to punish former Confederates and confiscate their lands. Instead, he ordered the return of confiscated and abandoned properties to those Southerners who met the conditions of his reconstruction plan. Additionally, the president took a lenient approach to granting pardons to those ex-Confederates who were required under his plan to make individual applications for forgiveness. By 1866, Johnson had granted more than 7,000 pardons.[283]

The End of a Way of Life

James Baxter, the leaseholder of the Hermitage, did not simply fade away at the end of the war. He insisted on remaining in possession of the estate until he was legally forced to leave. According to family records, to the last minute before the Bringiers were able to reclaim their home, family possessions continued to be carried away. On the very day, January 1, 1866, that the Bringiers were allowed to return home to the Hermitage, Baxter carried away much of the remaining livestock of the plantation as well as six beautiful watercolors of mythological figures which had been originally brought from Rome years before the war by Aglae's uncle, Bishop DuBourg. The last minute haul was so extensive that several plantation hands were commandeered to assist in moving the items to another of Baxter's properties in St. Landry Parish. Finding only the smoking debris of what was once the frames of the valuable pictures smoldering in the manor house's hearths, the Bringiers could tell that their cherished pictures had been hastily cut from their frames and spirited away with Baxter as he departed only a few hours before the arrival of Amedee and his wife Stella. Upon entering their home, which they had not entered for more than two years, the couple found the building stripped of many of the personal items they had been forced to leave behind. Included among the more memorable looted items was the family's fine collection of cut glass, china, toilet articles, a "pearl writing desk," many of Stella's wedding gifts, and all of the substantial collection of books from the library. According to Stella, "All was desolation and ruin on the place."[284]

The Bringier family's repossession of the Hermitage was only one step in their effort to regain a semblance of normality in their lives.[285] Though happy with their return to their home, they did not underestimate the magnitude of the task which lay ahead. Under even the most favorable of conditions, the production of sugar was a prodigious undertaking. Now with the damage and neglect caused by the war to their property and the region, including damaged equipment, buildings, lost laborers and livestock, neglected fields and levees, the Bringiers, like most other sugar planters, realized that if a recovery was to happen hard work by a dedicated and well-directed labor force would be essential. Yet the emancipation of the South's slaves had permanently changed the region's labor dynamics in such a way that many planters were unfamiliar with and unsure that the new wage-labor system could produce the reliable source of laborers needed to meet the rebuilding challenge. The difficulty experienced by planters in building a reliable postwar labor force continued to plaque the region's recovery efforts for years. For example, when Benjamin Tureaud attempted to address the labor issue by contracting on March 1, 1875, with 150 male hands to work the family's plantations in Ascension Parish, he experienced the frustrations of the new wage-labor system. Though both sides agreed to a daily wage of $1.00 without rations and with half of the wage being paid after grinding was completed and after all debts at the plantation store were satisfied, Tureaud's workers grew increasingly disenchanted with the work agreement, particularly the section that required them to fulfill the agreement "fully" or forfeit their back wages. The freedmen's unhappiness with the contract became critical in the midst of the grinding season when Tureaud complained in January 1876 that "Only three or four want to work in the field . . . and that it was with difficulty [he] could get enough to run the sugar house." Confronted with the realities of the new labor system, Tureaud was forced to increase his wages to $1.15 a day for men and $.85 for women. Similar disputes between workers and planters were common and led one Louisiana planter to bemoan that "the New System of Labor [was] controlled and guided . . . by Fanatics" and was even worse than many had feared. Faced with mounting frustrations and despair many planters lost hope and abandoned their efforts to rebuild.[286]

As with individuals across the South, the Bringiers faced the daunting challenge of postwar recovery with varying degrees of effort and success. After regaining possession of the Hermitage, Amedee and Stella attempted to retrieve as many of the personal items taken from the home during the war as possible. Fortunately, a friend of the family had successfully removed some of the family's items from the house for safekeeping. George Washington Graves was a longtime friend of Duncan Kenner. Before the war, he lived at Kenner's Ashland plantation and managed his extensive stable of thoroughbred horses. During the conflict, when Kenner was forced to abandon his estate, Graves obtained the lease of the property from the Federal government, saving Ashland from many of the problems and losses faced by Amedee and Stella at the Hermitage. As the lessee of Ashland and Kenner's Bowden property, Graves traveled about without interference from Federal authorities and as such was able to visit the Hermitage estate and somehow successfully removed some of the family's personal effects to Ashland. He returned these items to the Bringiers shortly after they reclaimed the plantation. As to the remaining missing items from the estate, Amedee and Stella attempted to recover them through legal action. The legal struggle to recover either the items or appropriate remuneration endured for decades. Family members pursued claims for generations. Although the family eventually obtained the personal support of several members of the United States House of Representatives in 1874 and again in 1887, the family failed to obtain satisfaction and persisted in a fruitless effort to press their claim for recovery well into the next century.[287]

The family's recovery efforts were not totally without success. In addition to the items saved by Graves, the family reclaimed several of their lost items through a strange stroke of luck. After spending years fruitlessly filing legal papers and pressing their claim against Federal authorities for the recovery of their missing personal items, a few of the prized items were eventually procured by the family in an ironic accident. Several years after the war, members of the family joined together in a business venture which included the acquisition of a warehouse in New Orleans. During their inspection of the property, they were surprised to find an old case in the building. Upon opening the shabby chest, they were astounded to discover a collection of books, which included several rare works that had been looted from the Hermitage during the war.[288]

The return to Hermitage gave the Bringiers hope that with effort their cherished family homestead could be returned to a profitable status and the pleasant homestead it once had been. Unfortunately, the challenges were more difficult to surmount than most expected. Such problems were not limited to the Bringiers. Adjustment to the postwar social order required Southerners to face the facts of postwar life in the South and forced them to abandon lifelong beliefs, habits, and prejudices. Such changes and challenges were more than some could handle. Many individuals and families, who in the ordered society of the antebellum years had successfully managed plantations for decades, experienced in the postwar years a gradual disintegration in their determination and character. The despair and frustration of the South's defeat and the inability of many planters to regain their prewar financial and social status after the war drove many Southerners to alcohol, gambling, idle amusements, and family quarrels. The depressed mood of the planters' war was aptly noted by historian J. Carlyle Sitterson in his history of the sugar industry when he wrote that, "The collapse of the economy left the majority of the people drifting without aim, without interest, without hope."[289]

The depressed condition of the postwar sugar industry added to the dire nature of the postwar recovery effort of the state's sugar planters. The war severely damaged the sugar industry, particularly in the areas of labor supply and machinery. In spite of determined efforts by many to rebuild, there was little left to build with and in the end little reward for the effort. Although sugar production had peaked in 1861 at 264,000 tons, by the war's end in 1865 production dropped to one of the smallest yields in history—9,950 tons. The industry was so damaged that it took until the final years of the nineteenth century for the sugar production totals to return to the output of the early war years. Devastatingly low prices further aggravated the situation. For those planters who did not simply give up and who attempted to rebuild, weak pricing continued to afflict the industry. Dropping from a 1864 peak price of 18 cents per pound, the price of sugar continued to slide to a 1880 value of only 6.5 cents and to less than 4 cents per pound in the 1890s.[290]

Like their neighbors, the Bringiers were faced with huge obstacles in their postwar endeavors to return their plantations to profitability. The financial recovery efforts of the family met with varying degrees of success. For family members who actively participated in the war on the side of the Confederacy, including Amedee Bringier, Richard Taylor, and Duncan Kenner, it took considerable energy and time for them to reacquire their property from the federal authorities. However, Aglae and Doradou's eldest son M. S., who was forty-seven years old when the war started, did not take an active part in the conflict. He remained home during the war and continued to keep and manage his properties. Not only was he able to keep some operations going at the old White Hall, Houmas, and Brulé properties, he made enough profit to liquidate the large financial claims against his mother's holdings. Despite surviving under extremely trying conditions during the war, the postwar economic and political conditions and uncertainties proved to be too much for him to handle and M. S. became the first family member to confront the realities of the new postwar sugar economy in Louisiana. The coming imposition of reconstruction policies controlled by radical forces in the North, which followed the congressional elections of 1866 and the dramatic political changes the election results foretold for Louisiana and the rest of the South, led M. S. to make the difficult decision to liquidate his interests in his plantation properties. Other than the year, 1867, the exact date that he gave up control of his share of the family's plantations is not certain. A surviving notarized document shows that by mid-January 1870, ownership of both his Houmas plantation and the Hermitage was in the hands of his three sisters, Nanine Kenner, Louise Gordon, and Aglae Tureaud. Yet annually published reports compiled at the time by agricultural economists P. A. Chompomier and

Alcee Bouchereau on production data concerning the sugar crop list his brother-in-law Benjamin Tureaud as the owner, suggesting that the property could have been owned by the sisters and managed by Tureaud. Nevertheless, by the time ownership of Houmas was transferred to his sisters the value of the property was estimated at $190,000, a third of the $575,000 offered by John Burnside in 1858. After giving up the plantations, M. S. moved to New Orleans where he joined several business enterprises with his brother-in-law Martin Gordon, Jr. Most notable of the partnership's projects was the operation of a sugar refinery in New Orleans on Commerce (now Magazine) St., not far from the family's Melpomene estate. M. S. continued to reside in the Crescent City until his death on August 22, 1884.[291]

The postwar survival story of the younger brother of M. S. was somewhat different. Unlike his older sibling, Amedee Bringier enthusiastically served in the Confederate army during the war. In spite of the many difficulties confronting him and the region's other returning veterans of the Southern Cause, Amedee refused to surrender to the despair and financial difficulties that devastated the hopes and efforts of so many others. Like the majority of the returning former Confederates who lived in the region, Amedee turned to the family's agricultural ventures for his postwar employment. Focusing on the rebuilding of the Hermitage, he was faced with a series of problems that many considered insurmountable, including the loss of a large portion of the estate's equipment, a difficult labor situation, and historically low prices. The road to recovery for Amedee and his family was long and arduous and ultimately met with mixed results. Never being one to commit himself partially to a task, Amedee gave his all to his effort to return profitability to the Bringier's sugar business at the Hermitage. As was the case with so many other planters in the postwar years, huge liabilities quickly built up and saddled Amedee and his wife Stella with suffocating debt. The situation became so serious that they nearly lost the Hermitage in November 1866 and again in May 1867 when portions of the plantation were placed on the selling block at sheriff sales. Fortunately, Amedee's siblings, Louise (Mrs. Martin Gordon, Jr.), Marie Elizabeth (Mrs. Benjamin Tureaud), Nanine (Mrs. Duncan Kenner), and Dadou Bringier, were determined not to let the Bringier family home slip from the hands of the family. Joining together, the four siblings purchased the plantation, thus saving it from falling into the hands of debt holders.[292]

Besides his endeavors to produce a successful and profitable sugar crop at the Hermitage, Amedee also applied himself to finding ways to assist the industry as a whole in its postwar recovery effort. He and his brother-in-law Duncan Kenner were among the first sugar growers in the area to install in their mills a new hydraulic regulator designed by John McDonald. The device reduced the time their mills were down for repairs and increased the juice extraction of the cane crop. Among Amedee's most ambitious and unique contributions to the postwar sugar industry was his design and production of several improved farm implements, including a patent in 1882 for a tailboard spring for wagons and carts used in the production of sugar. More significantly, he obtained U.S. patents for two additional inventions, one in 1879 for the "Improvement in Back-Bands for Plow-Harness," which provided a speedier disconnect system for plow-harnesses, and another in 1884 for his "Bringier Pulverizing Cultivator." His cultivator included a unique design of six revolving discs designed to improve the cultivation process in the sugar fields. Though technical successes, there is no evidence the devices were a great financial success for Amedee.[293]

Unlike many of their neighbors and friends, over time Amedee and Stella were able to repair much of the physical damage caused by the war to their home. The emotional devastation and demoralization of the four years of conflict, separation, and occupation were harder to handle. Like most white Southerners, they accepted the political, economic, and social consequences of their loss. However, the emotional consequences of the defeat were much more difficult to handle. Unwilling to repent for the past, after having expended so much personal and financial sacrifice to save their former way of life, Amedee and Stella remained deeply embittered and depressed over the death of the Confederacy and the loss of their prewar lifestyle. The couple clearly demonstrated the depth of their hurt and anger when they named a son born shortly after the war Booth, in honor of Abraham Lincoln's assassin. Tragically the child, the couple's seventh, died in infancy.[294]

The depression and emotional scars caused by the war left Stella and Amedee with more than just the embittered feelings they held toward the North. The stress and frustration of postwar survival also placed a growing strain on the couple's relationship. Though the marriage between Stella and Amedee was extremely fruitful—giving birth to thirteen children—it was not without its problems. The couple had experienced difficulties in their relationship even before the stress of their postwar struggle. Amedee's austere and formal disposition contrasted with Stella's impulsive and demonstrative personality and resulted in her having insecure feelings because of a nagging presumption that she had lost her husband's love. With the long separations resulting from Amedee's military service, her marital concerns only worsened. On at least one occasion during the war, Stella took the unusual step for a lady of the Old South and confided in a nephew, Bringier Trist, her concerns about her marriage. In January 1865, she took the even more unusual action of secretly writing her husband's commanding officer to plead that Amedee be granted a short leave of "two or three days" to be with her "to help sustain [her] fast failing courage."[295] Though somewhat unusual, Stella's wartime efforts to save her marriage paid off, at least temporarily, with the couple resolving their difficulties and remaining together for years. Following the end of the war seven of their thirteen children were born to the couple.[296]

Postwar life was not easy for the family. The financial, political, social, and personal problems that afflicted daily life in the postwar South placed an increasingly heavy strain on the family. Amedee's embittered feelings concerning the North only deepened as the years passed. He became increasingly austere and demanding in his dealings with his family. The estrangement in the family continued to mount until 1883 when a heated dispute between Amedee and his twenty-three-year old daughter Louise shattered the family. The situation evolved when the young lady fell in love with William Bateman. For unknown reasons, Amedee greatly disliked the young man. His efforts to dissuade Louise from her relationship with Bateman failed to discourage the young couple. Such insolence infuriated the former colonel who then blamed his wife Stella for plotting a scheme to circumvent his wishes.[297]

The matter exploded into full fury when Amedee learned through the errant remark by a relative that Louise and Bateman had publicly announced their marriage without his permission or knowledge. The former Confederate officer became so angry that he threatened to get his gun and kill his soon-to-be new son-in-law on sight. Cooler heads eventually prevailed upon Amedee, persuading him that such action would not be particularly wise. Relenting, the colonel instead declared that he would at a minimum assert his parental authority and required the young couple to leave the state and never again reside in Louisiana. Again he was advised that such demands on his daughter and her new husband were as empty as had been his admonitions about their marriage. Nearly bursting with frustration and anger over his inability to order obedience, Amedee proclaimed that as there seemed to be little he could do to influence the matter to his satisfaction, instead of forcing the young couple to leave the state, he would himself leave Louisiana and never reside there again. Furthermore, as he could not prove that his wife had not assisted Louise in her defiance of his authority, he would go without her![298]

Although Amedee's decision to desert his family and what he called his "miserable life" caused much distress for Stella and their children, the act was a long time in coming. The relationship between Amedee and Stella had been stormy for years. Within five years of their marriage, Amedee wrote his young twenty-one-year-old wife and scolded her for writing too many and too lengthy letters to him. In his note, written in the midst of Christmas week in 1855, he sternly wrote

> I have just received <u>three</u> letters from you and I hereby give you fair notice that if you write more than <u>two</u> letters per week I will not read any of them and will not write again—two sheets of paper and two stamps are more than enough. I cannot afford one and more. If you do not obey my <u>orders</u> I will knock off five dollars per month for post office expenses. (Your letters are too long, I have time to read only one page).[299]

The deteriorating nature of Amedee and Stella's relationship was again evidenced several years later when Amedee while on a trip to New Orleans cruelly taunted Stella in a letter with details of his meeting at the theater with a "charming, loving, bewitching" young lady who invited him to her room. After telling Stella of his meeting and invitation, he added

> Come be generous.... How can I resist such... the temptation is too great for a mortal being—I am going–going–gone–adieu.... Don't call me back too soon and promise that you will kick up no fuss If I give her about one third of my time. Once I get my little love... fixed up here—I'll go up to you my dear old wife and you'll find me only the better for you know that too much of a good thing is good for nothing—Variety is the spice of life.[300]

Even during the war, when it could be expected that long absences associated with his deployment with the Confederate army and the seriousness of Stella and the children's predicament of remaining at the plantation threatened by enemy troops, naval bombardment, slave discontent,

The End of a Way of Life

and lax law enforcement which endangered the safety of his family would have softened the relationship and his comments, Amedee continued his demanding tone and rebuked his wife for neglectfulness in the care of their children and even threatened her with divorce if she took the oath of allegiance to the United States.[301]

In spite of a desperate plea in which Stella proclaimed her undying love for him, Amedee was a man of his word and once he proclaimed in 1883 his intent to depart the state nothing she said or did altered his plans. He immediately began to liquidate his holdings in Louisiana by disposing of his property, including his share of the Hermitage. Fortunately, the estate did not pass entirely out of the hands of the family. Even before the calamitous consequences of the Civil War brought such hardship to the lives of the Bringiers, they had started to diversify the ownership of their plantations. In the decade before the war, Aglae had established a system of shared ownership of her plantations with two of her sons, M. S. and Amedee. After the war she gave her remaining share of the ownership of the Hermitage to her youngest son, Dadou. Thus when postwar financial problems at the Hermitage reached a crisis with portions of the plantation going to sheriff's sales, three of Amedee's sisters and his brother Dadou joined together to save the plantation. During the postwar years as the economic and political consequences of the war and reconstruction unfolded, ownership of the Bringier plantations continued to evolve with the Bringiers developing sophisticated schemes to keep control of their estates. In May 1867, Amedee and Stella went to court and received a judicial decree in which they divided ownership of their property, including the Hermitage, so as to give Stella title. This action they hoped would protect the ownership of the plantation in case of potential unforeseen developments arising from the changing economics and politics of reconstruction and Amedee's service to the Confederacy.[302]

By the time that Amedee and Stella's marriage collapsed in 1883, most of the economic and political uncertainty of the postwar and reconstruction years had been resolved. The intricate ownership schemes that helped the family save their holdings during the turbulent early postwar period were phased out. At the time of their separation, Nanine Kenner held Amedee and Stella's mortgage on their share of the Hermitage. Since Duncan Kenner had acquired part ownership of the plantation by obtaining young Dadou Bringier's share of the property, the Kenners now assumed responsibility for the plantation. The addition of Kenner's wealth and expertise into the ownership and management of the estate provided an important influx of resources, which helped to solidify the status of the property. Meanwhile, as promised, Amedee left the Hermitage, his children, his wife, and Louisiana to take up residence in Florida. Joined by his nephew Nicholas Trist, a cousin, Henry Octave Colomb, and later by his son Adee (Amedee), he moved to the St. Cloud/Tampa region of Florida where he became involved in the rapidly expanding Florida sugar industry. In 1889 he was hired as the general manager of one of Florida's largest sugar operations, the St. Cloud Plantation. Under his leadership, the company was reorganized into the Florida Sugar Manufacturing Company and expanded the cane acreage cultivated in his adopted state. By 1892 his sugar operation produced nearly 80 percent of all the sugar produced in Florida that year.[303]

While residing in Florida, Amedee seems to have shed some of his austere and proper demeanor. His change in behavior was so radical that Stella complained about the frontier atmosphere Amedee was exposing their son to in Florida. Shortly after her son joined his father in Florida, she angrily complained in a letter to another family member that her son was being subjected by his father to a "fearful den of the lowest order, roughs & cutthroats, [and] women devoid of virtue and refinement." Amedee remained in Florida for the rest of his life. Dying on January 9, 1897, his remains were returned home and buried in the Bringier tomb at Donaldsonville. Stella moved to Virginia for a short time and later returned to spend most of her remaining years in New Orleans where she lived with her sister-in-law Nanine Kenner at her residence on Carondelet Street. She died on November 11, 1911, and was interred in the family's large Donaldsonville tomb near her husband.[304]

Kenner's acquisition of the Hermitage was only one of a series of property transfers among the Bringier family after the war. Like most members of the prewar planter elite, with the possible exception of the well-connected Martin Gordon, Jr., every member of the family suffered large economic losses as a consequence of the war. The weak economic conditions in the postwar South weighed heavily on the recovery efforts of the family members as well as many other sugar growers in Louisiana. As time passed, debts increased to a point where many planters either had to sell or abandon their homes and lands. Among the extended Bringier family, Nanine and Duncan Kenner were the most successful in overcoming their wartime financial losses. Despite the fact that their Ashland plantation was raided and confiscated by Federal forces during the war, the Kenners were fortunate that a lifelong friend, George Washington Graves, acquired the lease to their lands. Therefore,

The End of a Way of Life 131

Kenner was in a better position than most absent planters to influence what was happening on his estate during the period of Federal occupation. Additionally, Kenner's keen acumen for business and finance had led him to diversify his investments prior to the war and provided him and Nanine with important resources which proved invaluable in helping them survive during the economically trying postwar period. The Kenners' fiscal standing allowed them to provide extensive financial assistance to other family members. Their help enabled many of their once prosperous relatives to survive the lean postwar years. The family's ties and love of the land made it difficult for the Bringiers to part with their properties. In most cases, every effort was made to keep the plantation within the family. Over time, as members of the family gave up their property, they often turned to the wealthy Kenners for help. By the time of his death on July 3, 1887 at age seventy-four, Duncan Kenner owned multiple properties in New Orleans and five large plantations in Louisiana's sugar country—Ashland, Hermitage, Bowden, Houmas, and Hollywood.[305]

Kenner used his Bringier properties to broaden his crop selections. With the postwar sugar industry in disarray, he looked for ways to both diversify his crops and to reinvigorate the sugar industry. Kenner turned to the growing of rice as an alternative crop. Using his Ashland estate as the center of his rice production effort, he quickly met with financial success and soon expanded production of the grain crop to his Houmas and Hollywood plantations. His efforts were so successful that by 1886 Kenner was producing more rice on his estates alone than the entire state in 1865. With Ashland being the center of his rice growing effort, Kenner focused on sugarcane production at the Hermitage and Bowden plantations. He was among the first sugar growers in the state to realize that if the industry was to ever recover from the wartime disaster, major changes had to be made. Kenner took the lead in awakening the state's sugar planters to the need of working together to improve the industry by revamping their production methods and forming professional organizations and agencies. He was among the first planters to use the Rillieux double-effect pans, the McDonald hydraulic pressure regulator, and the portable railroads to transport cane to his mills. Additionally, his efforts to improve the industry led to the formation of the Louisiana Sugar Planters Association in 1877 and in 1885, the Louisiana Scientific Agricultural Association (LSAA). Both organizations honored Kenner by selecting him as their first presidents, positions he held until his death in 1887. As leader of these organizations, Kenner played a vital role in stressing the importance to the state's agricultural industry for accurate statistics, scientific experiments, and result testing as a means of improving their products. He demonstrated his dedication to the scientific study of crop production in 1885 when he won the placement of a major United States Department of Agriculture project to install an experimental diffusion battery at the Hermitage. Although construction of the project commenced at the plantation, the experiment never reached full fruition because the Department of Agriculture's contractor, who had the responsibility for making the machinery needed for the project, failed to follow the proper design specifications for the equipment.[306]

In the postwar years, Kenner was involved in many activities in addition to his agricultural endeavors. He remained active in Louisiana politics during Reconstruction, serving in the state Senate from 1866 to 1867, and again in 1878. While in the legislature, he lead the passage of postwar legislation that became known as the Black Code. To many whites in the state the code was commonsense legislation designed to bring order back to the plantation labor system. To Northerners the legislation appeared to be designed to circumvent the newly ratified 13th Amendment to the Constitution by forcing former bondsmen into a near-slave status on the plantation and a direct affront to the victorious North. The outrage was so great over the code that the issue became a major factor in justifying the implementation of Radical Reconstruction across the South. Although out of office during most of the Reconstruction period, Kenner remained active in the Democratic Party's attempts to regain political control of the state. He eventually played a pivotal role in the disputed presidential and gubernatorial elections of 1876. The electoral controversy surrounding the presidential election of that year resulted in the end of Radical control and the collapse of Republican power in Louisiana. In spite of his dedication to the postwar Democratic Party, Kenner managed to maintain cordial relations with individuals on both sides of the political fence. Republican Pres. Chester A. Arthur disregarded Kenner's Democratic Party ties and appointed him to the United States Tariff Commission in 1882. In addition to his political activities, Kenner continued his prewar passion with thoroughbred horses. Being one of the founders of the Louisiana Jockey Club, he served as its president until his death in 1887. Before his death, Kenner continued to participate actively in many other civic endeavors, including serving as chairman of the building committee for the World's Industrial and Cotton Centennial Exposition held in New Orleans in 1884-1885. Dying in 1911, his wife Nanine survived her husband by a little more than two decades. Both Duncan and Nanine were buried in the Bringier family tomb in Donaldsonville.[307]

Although Kenner's postwar accomplishments were impressive, he was not the only member of the Bringier family during the years after the war to remain active in both public service and planter activities. After surrendering his brigade at Natchitoches at the end of the war, Gen. Allen Thomas still owned his New Dalton plantation in St. Landry Parish, Louisiana, and his Dalton estate in Howard County, Maryland. Unfortunately, both estates were severely impacted by the war. His loss of slaves from New Dalton alone was valued at $60,000. In an attempt to save what he could after the war, he quickly sold his Maryland holdings and decided to concentrate his efforts on his Louisiana property. Unfortunately, the losses he occurred during the war were more than he could overcome. In May 1866, Thomas and his family began a long odyssey of moving from place to place, never finding a comfortable and permanent family homestead. Leaving behind their home at New Dalton plantation, he and his family moved to a new cotton-producing plantation named Blackwater, located in Tensas Parish on Lake St. Joseph, not far from the Vicksburg battlefield where Thomas fought many desperate engagements during the war. Leasing the estate for $3,000 per year, Thomas and his family remained in the area for only three years. Thereafter, they moved from North Louisiana to the New Hope plantation in Ascension Parish, not far from the holdings of several other Bringier family members. There he and his family remained for thirteen years. In 1881, the family moved yet again, this time to Baton Rouge where he served from 1882 to 1884 as an instructor of agriculture at Louisiana State University and Agricultural and Mechanical College and later as a member of the Board of Supervisors for the university. After another short stay of only three years, the family again moved. This time they left the urban setting of Baton Rouge to live again on a rural estate—Hollywood plantation in East Baton Rouge Parish.[308]

During the postwar years, Thomas remained a loyal and active member of Confederate veterans' groups including the Army of Tennessee. As a fervent loyalist of the Democratic Party, he twice served as a presidential elector, first in 1872 for Horace Greeley's and then in 1880 for Gen. Winfield S. Hancock's unsuccessful presidential bids. In 1875, he was recruited by party leaders to run for Congress from Louisiana's third district but declined for business reasons. He also served as a member of the Democratic Party's State Central Committee and when he lived in Ascension Parish, he served ten years as the president of the party's parish's executive committee. While managing Hollywood Plantation, Duncan Kenner introduced him to Democratic Pres. Grover Cleveland in Washington, D.C. Shortly thereafter, the president appointed him as Coiner of the United States Mint at New Orleans. His new job necessitated still another move for the much-traveled Thomas family. Moving to New Orleans to a home on Esplanade Avenue at the corner of Claiborne Avenue, just a few blocks up the broad tree-lined street from where the Mint was located, Thomas remained Coiner for six years from 1885 to 1891.[309]

After completing his tenure as Coiner, Thomas and his wife moved yet again. This time, because of Mrs. Thomas' failing health, they migrated from Louisiana to Florida. Settling near Kissimmee, he acquired Runnymeade, an estate not far from the home of his brother-in-law Amedee Bringier. After the Democrat Grover Cleveland regained the presidency, Thomas accepted a call to national service when he was appointed in 1895 as United States Consul at La Guayra, Venezuela. A year later, he was promoted to the position of Envoy Extraordinary and Minister Plenipotentiary of the United States for Venezuela. While serving in this position, he played a pivotal role in settling a long-standing boundary dispute between Venezuela and Great Britain's British Guiana colony. Upon a request from Venezuela, the Cleveland administration became actively involved in the squabble on Venezuela's side. Much was made in both nations over the acrimonious nature of the dispute between the United States and Great Britain before an arbitration commission settled the issue. On his retirement from his post in 1897, the Venezuelan government honored Thomas for his outstanding service with the Order of Bolivar (first class). At the time, it was reported that it was the only instance of the award being conferred upon a foreigner. Between the financial losses of the war and his years of government service, Thomas and his wife Octavie faced an old age with few financial resources. Remaining in Florida until shortly before his death in 1907, Thomas and Octavie sold their property in the Sunshine State and acquired Istrouma at Waveland, Mississippi, on the Gulf Coast, a short distance from New Orleans. Desiring to be close to their family and friends in their final years, the couple had barely settled in their new home when Thomas took ill with malaria. Growing gradually weaker, the old general died on December 3, 1907. His wife survived him by nearly a decade. Both Allen and Octavie Thomas were buried in the Bringier family tomb at Donaldsonville.[310]

Of all the members of the family, none obtained a greater public profile during the war than Richard Taylor. His military skills won him admiration by individuals in both the South and the North. Unfortunately, Taylor's natural military prowess did not extend to the business world. Even before they were faced with dealing with their postwar economic troubles, Taylor and his wife Myrthe had struggled for sometime to keep their heads above the deepening economic waters. In the years prior to the outbreak of the war, when others planters and family members were accumulating huge fortunes, the Taylors borrowed large sums of money to keep their Fashion estate in St. Charles Parish in operation. Turning to Myrthe's mother, Aglae, for assistance, by the time Taylor left to offer his services to the Confederacy, he and Myrthe owed the granddame of the family a sum totaling in excess of $300,000. The events of the war years only worsened the Taylors' financial situation. When he returned home at the end of the war, the couple took the unusual step of not reclaiming their Fashion plantation. Although they had a legal claim for the return of their property, they did not have the resources to rebuild the sugar operations on the property. Instead, they retained their life right to the estate and allowed the government to continue to lease it to others. As their financial situation continued to worsen, the Taylors eventually found it impossible to repay their mounting debts. In 1866, in an attempt to find financial relief, the Taylors filed for bankruptcy in federal district court. Fortunately for the couple, neither Aglae Bringier nor Martin Gordon, Jr., who the Taylors also owed a sizeable sum of money, made any claim to their Fashion estate, which allowed them to retain their life-rights to the property.[311]

Not having either the interest or resources to return to the growing of sugar after the war, Taylor looked for other opportunities to restore his financial position. In 1866, the state opened the operation of the New Basin Canal—a man-made waterway linking the center of New Orleans with Lake Pontchartrain—for a public lease. Taylor successfully beat out the competition for the project and secured a fifteen-year lease on the waterway. In spite of the fact that the lease appeared to be a financial godsend, the former Confederate hero's business skills were not up to the task. Unable to meet even his first year's payment obligation to the state, Taylor again had to turn to a family member for assistance. Offering his brother-in-law Duncan Kenner a partnership in the project in January 1867, Taylor borrowed $100,000 from him for the project. In spite of the influx of new money into the undertaking, Taylor was unable throughout the entire life of the lease to pay the state even a single dollar for use of the canal. With the financial problems concerning his canal lease growing progressively worse, in August 1873 the state annulled the agreement with Taylor and later obtained a legal judgment against him, which the state was never able to collect.[312]

In spite of substantial financial difficulties, Taylor did not abandon his expensive lifestyle or his status as an aristocratic socialite. Along with Kenner, he was involved in the reorganization and management of the prestigious Metairie Jockey Club and the thoroughbred horse races it organized at its Metairie track. He also remained a member in good standing in the state's premier social organization, the Boston Club. Taylor also spent much time involved in postwar politics. Spending considerable time in the North after the war ended, Taylor used his social and political connections to secure the release of imprisoned Confederates, including Jefferson Davis. He also became involved in Reconstruction politics in Louisiana. However, Taylor suffered a severe loss of political prestige and influence in the state when Pres. Ulysses S. Grant backed out of a deal that Taylor claimed he had agreed to with him concerning the power struggle in Reconstruction Louisiana between Henry Clay Warmoth and William Pitt Kellogg. Leaving Louisiana to serve as European agent for a wealthy northern friend, Samuel Barlow, Taylor eventually returned to the state after a two-year absence. Shortly thereafter, his forty-one-year-old wife Mimi succumbed to a fatal fever. Devastated by his loss, he again left Louisiana to take up residence near his sister's home in Winchester, Virginia. There he continued his work as a representative and lobbyist for Barlow. In the controversial presidential election of 1876, Taylor returned to the political arena acting as an advisor to the unsuccessful Democratic nominee for president, Samuel J. Tilden. Disappointed once more with the controversial and contrived Democratic loss in the race, Taylor again forsook politics and returned his attention to more ordinary activities such as serving on the board of trustees of the Peabody Fund, which oversaw George Peabody's multimillion-dollar endowment for American education. He also spent time writing his memoirs on his wartime experiences. His research for his book brought him for a final visit to New Orleans and to the Bringier family's aging Melpomene home on Carondelet Street. His memoirs proved to be his final project. One week after the publication in 1879 of his highly praised *Destruction and Reconstruction*, while on a visit in New York City the fifty-three-year-old former general died of what was identified as dropsy—a condition of excessive fluid in the organs of the body usually resulting from heart disease and cardiac failure. His body was returned to New Orleans for burial in his tomb in

Metairie Cemetery alongside that of his beloved wife.[313]

Having grown close to his brother-in-law and commanding officer during the war, the youngest of Aglae's sons, Dadou, adopted many of Richard Taylor's cavalier and extravagant attitudes during the postwar period. Shortly after the end of hostilities, young Dadou traveled with Taylor to New York City where they enjoyed the many attractions of the city. During this period, Taylor made several trips to Washington, D.C., to meet with Pres. Andrew Johnson and other national leaders to discuss the fate of former Confederate leaders including his former brother-in-law, Jefferson Davis. Dadou remained in New York and continued to revel in the social life of the glamourous big city. Several weeks later other family members visited Dadou, including his nephew M. S. Bringier, Jr., and later Allen Thomas. His brother-in-law, as well as his former commanding officer during the war, Thomas was astounded by the extravagant lifestyle the young Bringier enjoyed in New York. The former Confederate general was so upset that he scolded Dadou for his lavish spending and partying—totaling approximately $10,000—at a time when so many Southerners and family members were suffering from the effects of the war. Thomas also verbally took his former commanding officer Richard Taylor to task for not placing limits on the uncontrolled spending of their young in-law.[314]

Whether it was the scolding or simply the fact that he grew bored with the New York City scene, not long after General Thomas' visit, Dadou sailed for Europe where he eventually settled in Paris. Remaining there for an extended length of time, he soon continued the extravagant lifestyle he enjoyed in New York. Quickly running through his remaining funds, he drew upon the Bringier family's European funds held by the Paris brokerage of Borde & Cie until he also exhausted that supply of money. Dadou returned home to war-torn Louisiana. In spite of his history of extravagance, his mother welcomed him home with open arms.[315]

Despite the difficulties Amedee was having at the time in rebuilding the Hermitage and restoring it to profitability, Aglae held a special place in her heart for her youngest son and jealously protected his share of the family's resources. She was determined to see that Dadou remained financially secure. He received a share of the family's Houmas estate and Aglae gave him her half ownership of the family's esteemed Hermitage Plantation. However, life on the postbellum sugar estate under the austere management of his brother, former commanding officer, and half owner of the estate, was not to the fun-loving Dadou's liking. After a short time he disposed of his share of the Hermitage to his sister Nanine's husband Duncan Kenner. Preferring the lifestyle and business operations in the city, Dadou moved into Melpomene with his mother and bought into the Milliken commission-merchant firm, making it Milliken and Bringier. Although the firm continued to handle some of the commission business of the various Bringier properties, he soon grew dissatisfied with his new business arrangement. According to family records, Dadou's seigniorial lifestyle did not transfer well to the business world of his new company. Milliken and other staff members in the company soon became irritated with his aristocratic and unproductive office manner and the partnership soon dissolved.[316]

His business failures did not affect Dadou's lifestyle. His fondness for spending money, international travel, drink, and frivolous friends continued unrestrained. Family reminiscences recount that one banquet given by him was so extravagant that the table was arranged in such a way that the diners sat around a miniature lake of champagne—the expensive wine gushed from fountains that fed the lake. As a final touch of grandeur, small yachts, which bore serving dishes that contained assorted delicacies, floated on the champagne lake. Such unbounded extravagance without a steady source of income soon resulted in the exhaustion of his, and much of his mother's, finances. Family accounts reveal that by the time he used up the funds made available to him by Aglae, his lavish lifestyle had cost the family approximately $125,000. In his latter years following the death of his mother, Dadou became largely dependent upon a monthly allowance given to him by Duncan and Nanine Kenner. Never marrying, Dadou's unsettled and troubling lifestyle continued to the end of his life. Gradually slipping deeper into a state of depression, death tragically came to him by his own hand on his forty-fifth birthday on August 3, 1887, one month exactly after the death of his financial benefactor, Duncan Kenner.[317]

The war and its aftermath also had a dramatic effect upon Aglae Bringier's life. As a widow living alone, she could have stayed at her home in New Orleans and suffered relatively few consequences from the occupation of the city by Federal troops. Although her own opinions remained largely aloof from the politics of the day, she supported her children and their spouses in their political stands and could not escape the turmoil associated with the Southern Cause. Not knowing what the war and the future had in store for the people of New Orleans and Louisiana, she determined early in the conflict that she would not live under Union occupation.

Aglae was staying at the Hermitage when Farragut's fleet broke through the Confederate defenses that protected New Orleans and thus was spared the stress and strain of the traumatic and chaotic evacuation of the city. Later when her security at the plantation was threatened, she joined with her daughters and other family members and evacuated into Confederate held areas of the state. Fortunately for Aglae, little damage was done to her Melpomene home in New Orleans while she was away. Searched by troops on several occasions, the house and its contents remained largely unaffected by the Union occupation of the city. The presence of her son-in-law, Martin Gordon, Jr., on the property helped to protect the house. Gordon lived in a house on the same square with Melpomene and maintained friendly relations with Federal officials in the city, including Gen. Nathaniel Banks. Such connections undoubtedly helped to save Aglae's house and furnishings from the fate that befell the family's properties upriver from New Orleans.[318]

Like so many others in Louisiana the strain and inconvenience of moving multiple times during the war to remain in safe areas, along with the tension of dealing with the emotional issues associated with having so many family members away from home and involved in the deadly conflict, placed an ever increasing amount of stress on Aglae. Her daughter-in-law, Augustine Tureaud Bringier, noted in a September 23, 1863, letter to her husband of Aglae's depression when she wrote "Your mother passed the day with us. I saw the poor old lady's tears course down her cheeks time and again. She worries over her absent children, not knowing where they are now are or when she will ever hear from them."[319]

Even when Aglae moved to live with her daughter Nanine and her husband, Confederate Congressman Duncan Kenner, and their children in the secure community at Natchitoches, her spirit did not recover. As a result of deepening depression, the decision was made by family members for her to return home to Melpomene in New Orleans. Under the care of the Gordons, the house and its furnishings had remained safe from either military raids or vandals. After a two-year odyssey across Louisiana as a war refugee, the granddame of the Bringier family returned home to the Crescent City and the house she loved on April 23, 1864. All members of the family noticed the toll that the war and refugee statist had taken on her. The changes were so recognizable that her grandson, M. S. Bringier, Jr., wrote his father that "Bonny arrived on the *Black Hawk* with Uncle Bob (Martin Gordon, Jr.), Tiny (Mrs. Augustine Trist) and her children. . . . She is terribly changed."[320]

Like so many other southern women, the war was a heavy emotional burden for the family's matriarch. She was fortunate that in spite of the active roles they played in the military and political life of the Confederacy, all of her sons and sons-in law returned home in good physical health after the war. Aglae, however, suffered the doleful and personal loss of one of her grandsons, Julien Bringier Trist, in the battle of Murfreesboro, Tennessee, on December 31, 1862. In addition to the emotional and heart-rending loss of her grandson and the knowledge that the family's fortunes and estates had suffered so greatly at the hands of Federal troops, Aglae also faced her own mounting financial crisis at the end of the conflict. Although she possessed substantial real estate holdings, like most members of her class, Aglae's prewar income was largely derived from sugar plantations. The neglect and devastation caused during the war converted the once profitable Hermitage and other family estates from being prewar lucrative ventures to postwar financial liabilities. Adding to Aglae's postwar financial plight was the seemingly uncontrolled spending spree of her youngest child Dadou. Her deep concern over his reckless demeanor in the months following the war and a desire to give him a more stable lifestyle led Aglae to turn over her remaining half-ownership of the Hermitage plantation to him. Furthermore, the war's devastating effect on the region's economy also limited the ability of other individuals, most particularly her son-in-law Richard Taylor, from repaying prewar debts owed to her. All of these issues resulted in Aglae's prewar fortune slipping away. The proud granddame of the family was gradually forced to sell most of the real estate holdings that she and her husband had acquired in the decades before the war.[321]

The weight of her financial woes eventually reached a point where her creditors took legal action against her, even threatening to seize her beloved Melpomene home and its furnishings. Fortunately for her, family members were determined to save the proper old lady from the anguish of destitution. When the sheriff threatened to seize the furniture and movables for nonpayment of her debts to other creditors, several family members developed an elaborate strategy to protect her holdings. Led by her daughter and son-in-law Nanine and Duncan Kenner, the family not only successfully saved her property from seizure but did so without Aglae ever knowing how close she had come to destitution and the true state of her financial affairs. More than any other members of the family, the Kenners survived the war with sufficient resources and ability to prosper in Louisiana's postwar economy. In the

The End of a Way of Life

years following the war, Duncan lent considerable funds to many family members including Aglae. Though generous with his loans, Duncan was a wise enough businessman who, whenever possible, secured his and Nanine's loans. Not even Mrs. Bringier was exempted from this business practice. For loans exceeding $60,000, Kenner obtained mortgages on Aglae's Melpomene home and furnishings in New Orleans as well as liens on her share of future crop production from her plantations.[322]

In spite of the businesslike way they handled their financial dealings with Mrs. Bringier, Duncan and Nanine were always careful to keep from her the full extent of her financial problems. Fortunately for the Kenners and the other Bringier family members, their efforts to keep Aglae in the dark about her financial misfortune were aided by her long-held custom of spending much of the hot summer months in Bay St. Louis, Mississippi. While there, she remained completely unaware of the events in New Orleans. As the sheriff prepared to seize her property and furniture in New Orleans, Duncan Kenner quickly moved to handle the issue. He immediately paid the creditor who was forcing the legal seizure the full amount owed and with the advice of family members, he shrewdly decided to let the sheriff's sale proceed. As Mrs. Bringier had used her belongings for collateral on other debts and had other judgments pending, future seizure attempts were likely unless the items were sold at a sheriff's sale with all of her creditors then receiving a share of the proceeds. At the sale Kenner simply outbid all other bidders and purchased all of the items himself. The items were returned to Melpomene as if they had been undisturbed and the grand matriarch never learned of the incident. The Kenners were so careful to protect Mrs. Bringier from the unpleasant realities of her financial condition, that in addition to securing her movable property, they also continued to pay without her knowledge her monthly allowance, which was supposed to have been guaranteed to her at the disposition of her husband's estate. The concealed payments by Duncan and Nanine continued to be made until the time of Aglae's death on June 11, 1878. As with the other members of the immediate family, interment was at the family's tomb in Donaldsonville.[323]

However, as with the South itself, the devastating effects of the war were permanent to the Bringiers' fortune and spirit. Financially, the loss of their wealth in slaves, equipment, and crops was more than they could work through. In spite of considerable effort by the family to stabilize the family's finances, with the exception of Duncan and Nanine

Kenner's holdings, prosperity never returned to the Bringiers and their plantations. In the end, all efforts by the Bringiers to restore their estates to their prewar prosperity failed and by the turn of the century nearly all of the family's extensive property holdings had passed out of their hands.[324]

Chapter 10
Epilogue

As the generation of Bringiers who grew up during the antebellum years aged and the economic and political vicissitudes of producing and processing sugar crops in postwar Louisiana increasingly wore on both the resources and temperament of the surviving family members, one by one the plantations were sold to new owners. After seven decades as the centerpiece of the Bringier family's plantation holdings and home to multiple generations of the family, the Hermitage passed from the hands of the Bringiers in 1888, following the death of Duncan Kenner. The sale of the property was more than a simple real-estate transaction, for it marked both the practical and symbolic end of a way of life for the Bringier family. The economic and political consequences of the Civil War and Reconstruction changed life in Louisiana's sugar country forever. In spite of the fact that the end of Reconstruction in Louisiana saw white elites return to political power, the economic challenges that the war brought to the sugar region of the state were so catastrophic that, like the Bringiers, many other sugar elites never reacquired either the financial or political status they had enjoyed before the conflict. In 1877, the year that Reconstruction ended in Louisiana, the sugar plantations in the state managed to produce only one-third of the output of the sugar that was produced in the sugar parishes during the peak years of the antebellum period. Reasons why the sugar planters had such difficult times after the war varied. Sugar production was an expensive undertaking that included both the growing and processing of the product which required not only agricultural and engineering prowess but also significant labor-management skills, particularly in the years following the abolishment of slavery.[325]

Contributing to the lower production rates of the sugar planters was a desperate shortage of capital. Because of the complicated and expensive production process for sugar, successful planters had always needed ready capital. By the end of Reconstruction, property values in the state experienced a 37 percent decline from prewar levels. The Bringiers and other individuals who relied heavily upon agricultural activities for their income were faced with the double burden of lower production totals and the loss of value in their land and livestock. By 1880 the value of farmland and related buildings declined more than 70 percent and farm livestock by 50 percent from their prewar worth. Planters also had the unfamiliar task of working with a paid labor force of field hands. Though most sugar planters attempted to replace their prewar slaves with poorly paid black work gangs, the effort was neither efficient nor easy. The increased racial tensions that accompanied the intimidation efforts of the elites in the state to reestablish white political control after the war only exasperated labor and race relations between sugar planters and their workers. The challenges were so difficult that approximately 30 percent of the cane fields in Louisiana that had produced successful crops prior to the war were left fallow during the uncertain postwar years. Unfortunately for many members of the prewar sugar-grower elite, the multiple calamities resulting from the war, including the loss of equipment and livestock, devaluation of their property, labor problems, and the increased racial tensions of the times forced many sugar planters to give up their agricultural pursuits. One antebellum planter in 1877 complained in a letter to the New Orleans paper the *Daily Picayune* that it was his personal experience that since the end of the war he had witnessed the ownership of nine-tenths of Louisiana's sugar estates change hands.[326]

So it was with the Bringiers. Though the family was able to keep most of their properties in the hands of family members for a time, by the end of the century the Bringiers no longer owned most of the family's large sugar-producing plantations. Included among those numbers of changed ownership was the sale of the family's beloved Hermitage. Although only one of several plantations owned by the family, the Hermitage, along with their city home Melpomene, had always held a special place in the hearts of members of the family. It was the homestead of Doradou and Aglae and as such was the centerpiece of family activities for decades. Its passing from the hands of the family not only marked a change in the Bringiers' lifestyle, but also was symbolic of the postwar transition of life and culture in Louisiana's sugar country. Though the postwar changes were dramatic, they did not necessarily result in an end to the sugar economy of the region. Many of the prewar sugar estates remained largely intact for sometime after the war. By using paid black field hands to perform the backbreaking physical labor involved in the sugar making process that formerly was performed by slaves, some sugar plantations were able to operate and produce a crop as they had before the war. Yet despite the efforts of many planters to

Epilogue

recreate as close as possible the prewar ways of the plantation, the war and Reconstruction brought conditions to Louisiana's sugar country which led to major political, economic, and social changes. Among the most notable of the changes experienced by the postwar sugar industry was the transition in ownership from the native oligarchical white-elite families of the prewar era to Northern entrepreneurs who flocked with capital in hand to the area after the war to buy up the financially-troubled estates from impoverished planters of the postwar period. Thus was the fate of several Bringier properties including the Hermitage.[327]

As the burdens of postwar-plantation ownership weighed upon family members, and the Bringiers began to entertain the sale of their properties, determined efforts were made to keep the plantations in the hands of the family. As noted earlier, when M. S. Bringier decided to give up his control of the Houmas/Brulé plantations, the property was transferred to the management of Benjamin Tureaud, his brother-in-law. Tureaud himself decided to give up ownership of his Union Plantation in St. James Parish during the same period. Although Tureaud produced sizeable crops on the Houmas/Brulé properties, in 1881 Duncan Kenner joined with him in a partnership in running the estate. With Tureaud's death in December 1883, his brother-in-law Kenner assumed sole control of the plantation until his death in 1887. From that time on, Houmas/Brulé no longer appeared on the list of active sugar-producing plantations. Houmas was not the only Bringier property that Kenner temporarily saved. Following the death of another brother-in-law, Hore Browse Trist, Kenner acquired ownership of Bowden plantation in Ascension, which he also operated until his death. Similarly, as previously discussed, Kenner actively worked to keep the Hermitage from falling out of the family's control. After acquiring young Dadou Bringier's share of the ownership of the plantation and partnering with Amedee to keep the plantation in operation, Kenner assumed ownership of the property when Amedee left Louisiana. Kenner continued to operate his and the Bringier plantations until his passing. As he aged, he brought in as a partner his son-in-law former Confederate general Joseph Lancaster Brent. However, with Kenner's death in 1887, Brent started to liquidate ownership of the properties.[328]

In the decades that followed the transfer of the estate from the ownership of the Bringiers, the property was acquired and sold by a series of owners. Purchased by W. D. Maginnis in 1888, the property remained under his ownership and continued to be used as an agricultural venture until early into the new century when the manor house and its surrounding lands were acquired by developers who proposed an ambitious real-estate development plan for the property. Edwin P. Brady of the Alluvial Land Purchases Company, Inc. in 1911 paid $75,000 for both the Hermitage and nearby Bocage plantation. Renamed the "New Hermitage" homestead, Brady and his associates put together a real-estate plan which called for the construction of nearly one hundred small farms on the development, with the old sugarhouse being used as a canning factory to process the produce from the new farms. Once populated and settled with families, Brady's plan called for the development to be transferred into a new community, which was to be known as the City of St. Elmo.[329]

The plan called for the New Hermitage lots to range in size from a little more than three acres to nearly eight acres and ranged in price from $192 to $1,045. Monthly term payments ranged between $10 and $15. To promote the proposal, the developers put together an aggressive marketing plan which promised potential buyers not only the pastoral pleasures of land ownership and rural living but also a profitable business venture of operating a truck farm that would benefit from the high demand for fresh produce in New Orleans and Baton Rouge. To assist those who did not have experience or knowledge in truck farming, the company operated a demonstration farm on one of the lots and hired "an expert agriculturist" who was available "without charge" to provide "advice and counsel" to the New Hermitage property owner. Additionally, the developers proposed a rather unusual marketing ploy for the time of niche marketing. Designed specifically to attract potential women buyers, the developers included in their promotional brochure a special section titled "A Word to the Wife." In the brochure the company avowed that, "The prospect shows clear to the mind's eye and it is not hard to imagine a cozy Farm Home with plenty of fresh eggs, fresh milk, fresh fruits, and fresh vegetables at hand. . . ." Continuing, the pamphlet promised that "Wives will welcome such an existence away from the strife and turmoil of city life, where daily bread, social ambitions, and expenses of keeping up with your neighbor in dress and fashion are struggles really not worth the effort that make them necessary." Centering their female customer pitch around the slogan "Let's get acquainted," the New Hermitage marketers attempted to close the deal with the promise that "Truly the women so situated is not only a pretty picture to contemplate, but also a queen reigning over a realm of her own command." Despite the ambitious and flowery marketing scheme of the Alluvial Land Purchase Company, little ever became of the "New

Epilogue

Hermitage" development, much less the dreamed City of St. Elmo. The reality of a sixty-mile one-way railroad trip to the market in New Orleans apparently prevailed over the hyperbole of the promoters.[330]

Owned mostly by absentee landlords, over the next several decades the once elegant Hermitage house suffered severely from neglect. By the 1940s, the house lay uninhabited and was even used in 1958 as a plantation ruin set in the movie *The Long Hot Summer*. Fortunately, the historic old home was saved from the fate of collapse that so many of its sister structures across the South have endured. After nearly two decades of abandonment, the distinguished old house was spared from an inglorious death by decay in 1959 when Robert and Susan Judice purchased the property. At that time the once grand estate of Doradou and Aglae Bringier was reduced to only the manor house and twenty-five acres.[331] The elaborate sugar mill and all the outbuildings, which were once part of the plantation, were gone.[332]

Though structurally sound, the Judices found the 145-year-old building in a seriously dilapidated condition. The roof was missing its shingles and was covered only with worn red corrugated tin. Inside of the house several of the door panels were missing and covered with chicken wire. Most of the original mantles in the building were gone, having been replaced with either cast-iron or simple black wooden mantles. The flooring on the ground level had to be replaced. Even the large floor sills of the house were nearly rotted through because they were constructed directly upon the region's damp alluvial soil.[333]

Determined to return the old building to its historic grandeur, the Judices commenced a gradual and painstaking historic restoration and remodeling of the house. During restoration, they worked methodically to repair the dilapidated and damaged portions of the house. From the tin roof down to the rotten sub-flooring, the new owners used authentic materials whenever possible to restore the historic structure. Among the more dramatic restoration effort was the use of flooring from the attic of the Hermitage and the demolished Helvetia plantation in St. James Parish to repair the wooden floors on the building's lower level. The chicken-wire door panels in the house were painstakingly replaced with cypress ones salvaged from doors acquired from Riverton plantation. Perhaps the most spectacular of all restoration activities in the house was the refurbishing of the fireplaces and the replacement of the mantles with circa 1795 ones obtained from the Armant plantation located near Vacherie.[334]

With the building's restoration well underway, the Hermitage was placed on the National Register of Historic Places in 1973. Today, the restored house sits back several hundred yards from the Mississippi River on Louisiana Highway 942 approximately two miles east of Darrow. Visitors approach the old manor house by traversing a long allée of young oak trees. Arriving at the house, visitors enjoy a premier example of one of Louisiana's dwindling number of antebellum plantation homes. Nearly two centuries after members of the Bringier family began construction on the Hermitage, it remains today a dynamic example of historic restoration, plantation architecture, and a living example of life as it was among Louisiana's elite families in old Louisiana. Today, the old house nestles among a collection of old cabins the owners have moved to the site and renovated. The Hermitage serves not only as a clarion example of historic preservation but also stands as the surviving symbol of the Bringier family and their contributions to the history and culture of Louisiana as influential members of the state's nineteenth-century ruling elite. However, just as significantly, the old house also survives as a lasting monument to the hundreds of individuals, including family members, slaves, and other workers, whose skills, sweat, and toil made possible the lifestyle and living conditions enjoyed by the Bringiers and other members of their class in times long passed.

Notes

Preface

1. Mary Ann Sternberg, *Along the River Road: Past and Present on Louisiana's Historic Byway* (Baton Rouge: Louisiana State University Press, 2001), 181.
2. The 141-year-old manor house at the estate was destroyed by fire on May 12, 2002.
3. Sternberg, *Along the River Road*, 174-175.
4. For the biography of Richard Taylor see T. Michael Parrish, *Richard Taylor: Soldier Prince of Dixie* (Chapel Hill: University of North Carolina Press, 1992) and for the biography of Duncan Kenner see Craig A. Bauer, *A Leader Among Peers: The Life and Times of Duncan Farrar Kenner* (Lafayette, La.: Center for Louisiana Studies, 1993).
5. T. Harry Williams, *Huey Long* (New York: Alfred A. Knopf, 1969), 456; William Ivy Hair, *The Kingfish and His Realm: The Life and Times of Huey P. Long*, paperback edition (Baton Rouge: Louisiana State University Press, 1996), 197.

Chapter One

6. Bringier Papers, IV:38, Hermitage Foundation Papers. References to the Bringier and DuBourg papers relate to a multi-volume typescript prepared by Rose Warren under the auspices of Robert Judice of the papers of Trist Wood. From the collection, four volumes of Bringier Papers and two volumes of DuBourg Papers were used in the preparation of this work. Archived copies of the material are part of the larger Hermitage Foundation Papers, which—at the time of publication—are held under restricted access at the Williams Research Center of The Historic New Orleans Collection and by the Hermitage Foundation in Darrow, La.
7. Because of the family's association with their native estate, the Bringier family, both in Europe and later America, was sometimes referred to as "Bringier de Lacadiere," Bringier Papers, I:61, 65.
8. It was about the time of their business problems that Vincent was lost at sea. It is unclear whether the business failure was the result of Vincent's loss or problems between the brothers. Despite the fact that members of the family believed that Vincent had died in a shipwreck at sea, some fifty years following the death of Marius, a stranger claiming to be the grandson of Vincent called upon the Bringiers. He claimed that Vincent had survived the wreck and had been taken to Scotland where he resettled. As the reason for this long-lost relative's visit to the Bringiers was to ask for financial aid to assist him in his return to his home, after having been stranded in New Orleans, some family

members questioned the validity of the stranger's story. After receiving the requested funds, he was never heard from again. See Bringier Papers, I:62-64a.

9. Bringier Papers, I:61a-62, 94; *New Orleans Picayune*, June 27, 1909.

10. Bringier Papers, I:65a-68.

11. Augustin was the son of the Bringiers' coachman. Born at the family's White Hall plantation, it is likely that during his youth he acquired the details of his story from older slaves. The story was told to Aurora Mather Tureaud who married one of Doradou Bringier's nephews.

12. Bringier Papers, I:69-69a.

13. The five plots included the Saumnier plantation acquired on October 13, 1785, the Boudreaux plantation on October 14, 1785, the Antailla plantation on June 17, 1787, and two separate plots known as the Melancon plantations, of which the records of the date of one plot have been lost and the other was obtained on March 1, 1787. See Bringier Papers, I:72.

14. Bringier Papers, I:73-73a.

15. Paul Malone and Lee Malone, *Louisiana Plantation Homes: A Return to Splendor* (Gretna, La.: Pelican Publishing Company, 1996), 7; Lloyd Vogt, *New Orleans Houses: A House-Watcher's Guide* (Gretna, La.: Pelican Publishing Company, 1992), 15.

16. There remains some mystery about the actual outer-layer of the house. Although family records and tradition maintain that the house was covered with a marble veneer, at the time the building was put up for sale a detailed advertisement for the property described the structure simply as "a two story brick house." On the other hand, visitors to the site of the old home in the years following its destruction often mused about the large amount of marble fragments they found scattered around the site. These fragments could have led family members to conclude that the house was covered with marble. See *Gazette de la Louisiane*, September 1820 and Bringier Papers, IV:47.

17. *New Orleans Picayune*, June 27, 1909; Bringier Papers, I:69-69a; Bringier Papers, IV:43-44a; Fred Daspit, *Louisiana Architecture 1714-1830* (Lafayette, La.: Center for Louisiana Studies, 1996), 197-198; William Nathaniel Banks, "The River Road Plantations of Louisiana," *Antiques* 3 (June 1977): 1176-1177.

18. Pierre Clement de Laussat, *Memoirs of My Life*, ed. Robert D. Bush, trans. Sister Agnes-Josephine Pastwa (Baton Rouge: Louisiana State University Press, 1958), 66.

19. Ibid.

20. The arpent was a French measurement of land that varied somewhat from locality to locality. Usually lineal, it typically measured .84 to 1.28 acres and referred to the length of one side of a square. See J. Carlyle Sitterson, *Sugar Country: The Cane Sugar Industry in the South, 1753-1950* (Lexington: University of Kentucky, 1953), 11.

21. *Gazette de la Louisiane*, September 1820.

22. Ibid.; *New Orleans Picayune*, June 27, 1909; Bringier Papers, I:74; Zette

Trudeau and Felicie Bringier to Trist Wood, October 27, (1907?) in Bringier Papers, IV:50-51.

23. Bringier Papers, I:75 and 85; *Gazette de la Louisiane*, September 1820.

24. Bringier Papers, I:74a.

25. Though Trist Wood and others speculate that while in the colonial militia Marius participated in Spain's campaigns against the English during the American Revolution, in actuality the conflict had ended by the time he enlisted. Bringier Papers, I:70-71, 85.

26. Bringier Papers, I:74a and 76; *New Orleans Picayune*, June 27, 1909.

27. Because of Marius Pons' death on April 23 of the same year, 1820, that Audubon first arrived in Louisiana, there is some question as to whether Audubon actually visited the Bringier home. However, considering Audubon's love of birds and the Bringiers' reputation for hospitality, it would not have been that unusual for Audubon to have been drawn to the estate where Marius Pons kept his celebrated collection of birds.

28. Bringier Papers, I:74a and 76; *New Orleans Picayune*, June 27, 1909; Lillian C. Bourgeois, *Cabanocey: The History, Customs and Folklore of St. James Parish* (New Orleans: Pelican Publishing Company, 1957), 139.

29. Bringier Papers, I:83; *New Orleans Picayune*, June 27, 1909; Joseph G. Tregle, Jr., *Louisiana in the Age of Jackson: A Clash of Cultures and Personalities* (Baton Rouge: Louisiana State University Press, 1999), 130, 132; Robert V. Remini, *The Battle of New Orleans: Andrew Jackson and America's First Military Victory* (New York: Viking, 1999), 186-187, 194-195; Roger Baudier, *The Catholic Church in Louisiana* (New Orleans: Roger Baudier, 1939; reprint Louisiana Library Association, 1972), 265.

30. Included among the children were sons Paul Louis (1784-1860, known as Louis) and Michel Doradou and daughters Francoise (1786-1827, wife of Christophe Colomb, known as Fanny), Louise Elizabeth (1788-1863, known as Betsy and wife of Augustin Dominique Tureaud), Francoise Laure (b. 1792, known as Laure and wife of a Mr. Baron), and Melanie Elizabeth (1793-1863 and wife of a Mr. Wilson). See Glenn R. Conrad, ed. *A Dictionary of Louisiana Biography* (New Orleans: Louisiana Historical Association, 1988), s.v. "Marius Pons Bringier," by Florence M. Jumonville.

31. Bringier Papers, I:89a, 94.

32. Ibid., 99.

33. Ibid., 85, 200.

34. Ibid., 85-86.

35. Ibid., 87.

36. Ibid., 88.

37. Sitterson, *Sugar Country*, 24; Bringier Papers, I:88; Bringier Papers, IV:137.

38. Sitterson, *Sugar Country*, 45-47, Bringier Papers, III:60a; Bringier Papers, IV:45-46.

Chapter Two

39. Over time there has been much debate in Louisiana over the word "Creole." At different times and locations in Louisiana's history the meaning of the word has evolved. Today, the term commonly refers to persons of mixed ethnic or racial heritage. However, at the time of the Bringiers, when Louisiana was experiencing a rapid influx of immigrants, the term applied to natives of Louisiana, regardless of their ethnic background. It is in this classic context that reference is made to Creole in this work. See Joseph G. Tregle, "On That World 'Creole' Again: A Note," *Louisiana History* 23 (Spring 1982): 193-198.

40. Bringier Papers, I:215-216; Robert Dabney Calhoun, "A History of Concordia Parish Louisiana," *Louisiana Historical Quarterly* 15 (January 1932): 53.

41. Bringier Papers, I:216-216a.

42. Ibid., 218-221. It is probable that it was the ill-fated gambling adventure that led to Louis' sale of his plantation to J. Tricou on March 5, 1807, and his father's re-purchase of the property two days later. It is possible that Louis sold the estate as a means of raising funds for his gambling debacle and his father's quick acquisition two days later was part of his effort to address the issue of the disappearance of his son and his cotton and indigo proceeds.

43. Ibid., 222-222a, 251-251a; Benjamin Silliman, "Letter From L. Bringier, Esq. Of Louisiana to Elias Cornelius," *The American Journal of Science and Arts* 3 (1821): 19-20; Henry Thomason, "Ancient History Interpreted at Toltec Mound," Arkansas Department of Parks and Tourism, October 1, 2002, http://www.arkansasmediaroom.com (accessed March 8, 2007).

44. Bringier Papers, I:225-225a.

45. American State Papers, House of Representatives, 12th Congress, 1st Session, Public Lands: Vol. 2, 366; Bringier Papers, I:226-227.

46. Bringier Papers, I:228-230.

47. Ibid., 231-232.

48. This was somewhat of an exaggeration. His only relationship to the bishop was that his brother was married to a niece of the prelate.

49. Bringier Papers, I:233-235.

50. Ibid., 236-237.

51. Ibid., 238, 241; Roulhac Toledano and Mary Louis Christovich, *New Orleans Architecture*, vol. 6, *Faubourg Treme and the Bayou Road* (Gretna, La.: Pelican Publishing Company, 1980), 29; Receipt dated September 16, 1837, for $1,987.50 paid by Doradou Bringier to Alexander Milne as partial payment for lands situated between the New Canal and Bayou St. John in Robert Judice Collection, Williams Research Center, The Historic New Orleans Collection, New Orleans, La. (hereafter cited as the Robert Judice Collection).

52. Though heavily mortgaged towards the end of his life, Don Louis lived in the house until his death. After the Civil War, the building was demolished to make way for the construction of several smaller homes on the space. Bringier Papers, I:242, 246.

53. Ibid., 238-238a, 251a.

54. Ibid., 238a.

55. Their children included: Letitia Bringier (born November 26, 1833), Louise Bringier (born August 20, 1835), and Charles Pendleton Bringier (born December 19, 1839).

56. Bringier Papers, I:239-241, 255-256a.

57. Ibid., 261.

58. It is most likely that the wedding gift of the plantation was similar to Marius Pons' deal with Doradou and was probably an advance on Fanny's inheritance.

59. Bringier Papers, I:261a-271.

60. Ibid., 264-266.

61. Ibid., 267-268.

62. Ibid., 274-275.

63. It should be noted that not everyone was as taken by Christophe's talents as was Fanny. When the French Commissioner and former Prefect visited White Hall in 1803, he found Marius Pons' son-in-law to be merely "a second-rate dauber in paints." Laussat, *Memoirs of My Life*, 64; Bringier Papers I:275-276.

64. At the time of their return, Bocage was managed by one of his and Fanny's two sons. See Bringier Papers, I:263, 274, 277.

65. A story unsupported with any documentation other than family recollections tells of a dispute and duel between Christophe and his brother-in-law Doradou Bringier. Upon his return to Bocage following his second marriage, there was considerable displeasure among members of the Bringier family for the way that Christophe was treating his children from his marriage with Fanny. The disagreement led to a challenge and then a duel between the two brothers-in-law. Pistols were the weapon of choice. Christophe fired first and missed altogether. Instead of returning fire, Doradou—a dead shot who was well known for his ability and accuracy in the use of firearms—reportedly told Christophe that he would reserve his privilege of shooting him for the next time he found out that his sister's children were being unfairly and unjustly treated by their father. Ibid., 263, 277, 279, 283, 288-289.

66. Bringier Papers, II:1.

67. Paul Lachance, "Marriage and Property in Antebellum New Orleans," (paper presented at annual conference of the Association of Caribbean Historians, Havana, Cuba, April 15, 1999), 4.

68. Quoted by Lachance, "Marriage and Property in Antebellum New Orleans," 5.

69. Ibid.; Bringier Papers, II:8-14.

70. Located up the Mississippi River from the family's White Hall estate, Union Plantation was a gift—probably an inheritance advance—by her father to his daughter and new husband at the time of their marriage. Bringier Papers, II:15-23.

71. Ibid., 115-123.

72. Ibid., 125-129; *United States v. Watkins*, 97 US 219 (1877).

Chapter Three

73. Grace King on page 415 of her 1921 work *Creole Families of New Orleans* (New York: MacMillan Co.) states that Doradou was born on his father's plantation. She presents no source to support her claim. Practically all other sources on this subject support Trist Wood's claim that he was born at sea.

74. Canon J. B. Bringier to Douradou Bringier, June 29, 1836, Robert Judice Collection; Bringier Papers, II:34-35a, 51; Bringier-Colomb Family in Louisiana 1752-1943 genealogical chart compiled by Amedee Colomb, Sr. and revised by Clifford Colomb, July 1943, provided to the author by Duke Rivet.

75. His full name was Pierre Francois DuBourg, Sieur Chevalier de Ste. Colombe.

76. DuBourg Papers, II:202-206; Grace King, *Creole Families*, 399.

77. DuBourg served as a major in the Louisiana Volunteers, Collector of the Port of New Orleans, and as Finance Minister for Gov. W. C. C. Claiborne.

78. DuBourg Papers, II:247-248, 262.

79. Her full name was Louise Elizabeth Aglae DuBourg de Ste. Colombe. DuBourg's sons-in-law included the *anciennes* Michel Doradou Bringier and John (Jean) Pierre Francois Thibaut and the Americans Horatio Davis, Seaman Field, and John Harvey Field. See DuBourg Papers, II:198a and King, *Creole Families*, 400.

80. L'Abbé DuBourg became one of the Catholic Church's leading clerics in America. While in Baltimore he served as President of Georgetown University and founded St. Mary's College. Appointed Apostolic Administrator in New Orleans, he would later be named Bishop of Louisiana and serve in that position at the time of the Battle of New Orleans. He later founded St. Louis College and eventually returned to France, where he spent his final years. King, *Creole Families*, 399.

81. Bringier Papers, II:38, Bringier Papers, IV:66; DuBourg Papers II:283.

82. Bringier Papers, II:85-85a.

83. Quoted in Annabelle M. Melville, *Louis William DuBourg: Bishop of Louisiana and the Floridas, Bishop of Montauban, and Archbishop of Bescancon, 1766-1818*, vol. 1, *Schoolman: 1766-1818* (Chicago: Loyola University Press, 1986), 164-165, also see 155-157.

84. Bringier Papers, II:105, and 80a, 82; Bauer, *Leader Among Peers*, 22.

85. DuBourg Papers, II:212; Bringier Papers, II:40-41, 83-84.

86. Typescript copy of prenuptial agreement, June 17, 1812, Bringier Papers, II:40-42.

87. Ibid., 86-87a; Catherine Clinton, *The Plantation Mistress: Women's World in the Old South* (New York: Pantheon Books, 1982) 60, 233; Lachance,

Notes for Pages 29 - 36

"Marriage and Property in Antebellum New Orleans," 1.

88. Bringier Papers, II:89.

89. At the time Doradou obtained the property on which he would later build his stately home, the parcel measured twenty arpents across the front by forty arpents deep. DuBourg Papers, II:212; Bringier Papers, II:37, 41a; Bringier Papers, IV:67-68, 118.

90. It is probable that Marius Pons' wedding gift to the young couple was similar to the inheritance advances he provided his other children at the time of their weddings. DuBourg Papers, II:212; Bringier Papers, II:37, 41a; Bringier Papers, IV:67-68, 118.

91. Bringier Papers, II:46.

92. Ibid.

93. Ibid., 57a-58a.

94. Ibid., 47.

95. Ibid., 60, 64.

96. Ibid., 60a.

97. Bauer, *Leader Among Peers*, 85; Bringier Papers, II:63-63a.

98. Bauer, *Leader Among Peers*, 82; Bringier Papers, II:64.

99. Baby Hyppolite Christophe died at age fifteen days and was buried on January 9, 1826.

100. DuBourg Papers, II:23-23b; Bringier Papers, IV:75, 85-85a; Bauer, *Leader Among Peers*, 21-22, 83.

101. As was the case with his father, the spelling of young Doradou's name was not always consistent. In addition to the more common Doradou, in family and legal records of the time his name sometimes appears as Douradou and even Dauradou.

102. Bauer, *Leader Among Peers*, 21-22; Bringier Papers, II:145; Bringier Papers, III:132, Bringier Papers, IV:22; Student notebooks, 1844 and 1847, Louis A. Bringier and Family Papers, Louisiana and Lower Mississippi Valley Collections, Hill Memorial Library, Louisiana State University, Baton Rouge, Louisiana (hereafter cited as Louis A. Bringer Papers).

103. Quoted in Bauer, *Leader Among Peers*, 90-91; Bringier Papers, I:19-20; Bringier Papers, II:95; Property deed for land on the Bay of St. Louis in Hancock County, Mississippi, to Aglae Bringier, April 28, 1854, photocopy provided to author by Robert Judice.

104. William Kauffman Scarborough, *Masters of the Big House: Elite Slaveholders of the Mid-Nineteenth-Century South* (Baton Rouge: Louisiana State University Press, 2003), 53-56.

105. M. S. Bringier to his mother Mrs. M. D. Bringier, May 19, 1849, photocopy provided to the author by Robert Judice.

106. H. B. Trist to N. P. Trist, April 3, 1847, Manuscript Department, University of Virginia, Charlottesville, photocopy provided to the author by Dr. Robert Judice.

107. Bringier Papers, II:64-67; See Appendix A for listing of tomb placements.

108. Bringier Papers, II:68, Bringier Paper, IV:137a-139; H. B. Trist to N. P. Trist, April 3, 1847.

109. Named as Marius Ste. Colomb on his baptismal documents, on legal documents of the time his name often appeared as "Ste. Colombe Marius Bringier. However, to friends and many family members he was known as M. S. or "Uncle MS." Finally, to his most dear family members, including his mother, he was referred to as "Bringier" or "Chere Bringier." Bringier Papers, II:144-144a.

110. Bringier Papers, IV:109-112, 137.

111. The question of M. S.'s ownership of Houmas plantation remains somewhat unsettled. Although his daughter Zazette told Trist Wood that her father never owned any part of the plantation and that for a time in 1850 he even gave up management of the property, there exists considerable evidence, such as census reports and other family documents, which suggest that he obtained at least partial ownership from his mother around the time that she gave partial ownership of the Hermitage to Amedee. Bringier Papers, II:146.

112. Bringier Papers, IV:71, 109-112; Bringier Papers, II:144-145.

113. Bringier Papers, II:149-152; See Appendix B for listing of Bringier family patents.

114. Bringier Papers, IV:111, 113; Th. H. Kennedy to Allen Thomas, May 10, 1858, Robert Judice Collection.

115. H. B. Trist to N. P. Trist, April 3, 1847; Clinton, *Plantation Mistress*, 76-78.

116. Bringier Papers, II:71, 79, 110, 148 and Bringier Papers, IV:137a-139; Parrish, *Richard Taylor*, 61; *New Orleans Daily Picayune*, June 11, 1878.

117. Quoted by Parrish in *Richard Taylor*, 61; Bringier Papers, III:68-69, and Bringier Papers, IV:40-40a.

Chapter Four

118. Oscar T. Barck, Jr. and Hugh T. Lefler, *Colonial America*, 2nd ed. (New York: The Macmillan Company, 1968), 283, 323.

119. Bauer, *Leader Among Peers*, 34-35; Clinton, *Plantation Mistress*, 18.

120. The Hermitage property in the fourteen-year period prior to Doradou's final land acquisition in 1812 actually had a total of seventeen different owners and consisted of two different land tracts owned by Marin Landry and Gregoire French. Landry's tract was approximately fourteen arpents wide and was successively owned by Landry, Git Leblanc (1798); M. Andry (1802); L. Bringier (1804); J. Tricou (1807); M. Bringier (1807); J. Mercier (1808); F. and H. Amelung (1811); Syndic of creditors (December 1811); and finally Michel Doradou Bringier (August 24, 1812). The second and larger tract of twenty arpents devolved from French to James Mather (March 17, 1791); William

Conway (March 17, 1791); Pierre Part (March 26, 1791); M. Bringier (May 7, 1804); J. C. Westerstrandt (May 7, 1804); M. Bringier (October 10, 1804); and finally, Michel Doradou Bringier (November 27, 1806). Bringier Papers, I:82, 218; Bringier Papers, II:41; Miscellaneous Notes Folder, 24, Hermitage Foundation Papers; Synopsis des titres de M. D. Bringier, 1833, photocopy provided to the author by Robert Judice.

121. Bringier Papers, I:80; Bringier Papers, IV:118; undated typescript of court document signed by M. Bringier and M. Doradou Bringier before Judge A. D. Tureaud in the possession of Robert Judice.

122. The title of "architect" had little legal or professional standing until fairly recently. During the antebellum period anyone wishing to do so could call themselves an architect. Hence, the term architect was seldom used in most of the Old South. The artisans who provided services to planters and others and assisted in the construction of great homes were usually called by terms which denoted their particularly skill, such as "carpenter" or "surveyor." See Bauer, *Leader Among Peers*, 35.

123. Sitterson, *Sugar Country*, 73; Bauer, *Leader Among Peers*, 37.

124. Among some architectural historians there is some debate over the original style of the Hermitage. Some believe that when it was first constructed, the building took the traditional colonial style of the period and was not an early example of the colonnaded Louisiana-Classical styling. They maintain that the Bringiers remodeled the structure later and updated its styling. However, restoration work by the current owner, Robert Judice, revealed that the columns and the style were original and dated to the early construction of the house. Richard Sexton, *Vestiges of Grandeur: The Plantation of Louisiana's River Road* (San Francisco: Chronicle Books, 1999), 20-21; Bauer, *Leader Among Peers*, 35; Banks, "River Road Plantations of Louisiana," 1179.

125. Bringier Papers, II:44.

126. Ibid.

127. Ibid., 43, Bringier Papers, IV:137; Hermitage Plantation Journal, 1833, Louis A. Bringier Papers, translated typescript copy in possession of Robert Judice.

128. The diminutive size of the interior layout of the Hermitage was not uncommon. Visitors to many of the region's old estates are surprised at the limited living space found within the walls. The grand columns and galleries seen from the outside give an impression of massiveness that was seldom found on the inside of the old structures. Though there are indeed exceptions to this trait, including Doradou's daughter Nanine and her husband Duncan F. Kenner's magnificent Ashland manor house located a few miles upriver from the Hermitage. See Banks, "River Road Plantations of Louisiana," 1170.

129. Today the Hermitage remains perhaps the earliest Greek Revival plantation home in Louisiana. See Daspit, *Louisiana Architecture*, 269-270; Banks, "River Road Plantations of Louisiana," 1173, 1179; Malone and Malone,

Louisiana Plantation Homes, 60; Robert Judice to the author, April 19, 2008.

130. Undated note from Robert Judice to the author.

131. Ibid.; Daspit, *Louisiana Architecture*, 269.

132. Banks, "River Road Plantations of Louisiana," 1173, 1179; Herman Boehm de Bachellé Seebold, *Old Louisiana Plantation Homes and Family Trees* (New Orleans: Herman Boehm de Bachellé Seebold, 1941), 135; Malone and Malone, *Louisiana Plantation Homes*, 60; Livre Contenans les affaires De M. Doradou Bringier, Louis A. Bringier Papers; undated notes from Robert Judice to the author.

133. DuBourg Papers, II:130; Jessie J. Poesch, "Furniture of the River Road Plantations in Louisiana," *Antiques* 3 (June 1977): 1184-1185.

134. Sexton, *Vestiges of Grandeur*, 82; Malone and Malone, *Louisiana Plantation Homes*, 60.

Chapter Five

135. Edwin Adams Davis, *Louisiana: A Narrative History*, 3rd ed. (Baton Rouge: Claitor's Publishing Division, 1971), 203; Charles L. Dufour, *Ten Flags In the Wind: The Story of Louisiana* (New York: Harper and Row, 1967), 151.

136. William H. Russell, *My Diary, North and South* (London: Bradley and Evans, 1863), 109.

137. Davis, *Louisiana*, 203; Sidney A. Marchand, *The Flight of a Century (1800-1900) in Ascension Parish Louisiana* (Donaldsonville, La.: n.p., 1936), 67; Lionel C. Durel, "Creole Civilization in Donaldsonville, 1850, According to 'Le Vigilant'," *Louisiana Historical Quarterly* 31 (1948): 988-991.

138. Photocopy of undated property inventory provided to the author by Robert Judice; William R. Mitchell, Jr., *Classic New Orleans* (New Orleans: Martin-St. Martin Publishing Company, 1993), 166.

139. Although reserved for the construction of a canal, the wide median also served as a convenient boundary between the American and French communities who were experiencing great difficulty in blending their languages, religions, and customs and soon was dubbed the "neutral ground" between the two groups. To this day, medians throughout the Crescent City and its surrounding regions continue to be known as neutral grounds. See Mitchell, *Classic New Orleans*, 104.

140. Mary Louis Christovich, Sally Kittredge Evans, and Roulhac Toledano, *New Orleans Architecture*, vol. 5, *The Esplanade Ridge* (Gretna, La.: Pelican Publishing Company, 1977), 98, 100. Although the Bringiers moved from the building, it remained in the family for a long time. Eventually sold by Doradou to his son-in-law Duncan Kenner for $30,000, the structure was held by the Kenner family until the turn of the century when his then aging wife Nanine sold the property in October 1905 for $160,000. By the time the old Bringier city residence passed out of the family's possession the property had been subdivided and was then located next to the Grand Opera House.

After passing out of the hands of the Kenners, the structure was leveled and the Audubon building was constructed on the site in 1909, which later became known as the Maison Blanche Annex. See Bringier Papers, IV:84.

141. Originally named Apollo Street, the name of the road was changed by the city in 1852 to Carondelet. At the same time the more prominent street of Naydes (sometimes Naides) was changed to St. Charles Street. Bringier Papers, II:55, Bringier Papers, IV:70, 93 and Bringier Papers, I:84.

142. John Kemp, *New Orleans: An Illustrated History* (Woodland Hills, Calif.: Windsor Publications, Inc. 1981), 78-79; Peirce F. Lewis, *New Orleans– The Making of an Urban Landscape* (Cambridge, Mass.: Ballinger Publishing Company, 1976), 38-39.

143. Field was married to Aglae's younger sister Eliza. See King, *Creole Families*, 400.

144. Seaman Field was suffering financial difficulties at the time so he offered for sale shares of Melpomene. Also owning a one-sixth share was Martin Gordon, Sr. Only a year before he obtained full ownership was Doradou able to increase his share of the estate to one-third interest. It remains unclear how Doradou and his family were able to move to the house when they did. Perhaps rent was paid to Field to help their kinsman through his financial problems? See Bringier Papers, IV:91.

145. Samuel Wilson, Jr., and Bernard Lemann, *New Orleans Architecture*, vol. 1, *The Lower Garden District* (Gretna, La.: Pelican Publishing Company, Inc., 1971), 27; Bringier Papers, IV:85, 91.

146. The last major flood to hit the area around Melpomene occurred as late as 1849 when the great Sauve crevasse inundated nearly all of the back areas of the city and even reached into portions of the Vieux Carré. Wilson and Lemann, *Lower Garden District*, 28; Bringier Papers, IV:85-89, 93; Vogt, *New Orleans Houses*, 33.

147. M. S. Bringier to his mother Mrs. M. D. Bringier, May 19, 1849, photocopy of letter provided to the author by Robert Judice.

148. Bringier Papers, I:2a; Bringier Papers, IV:106; DuBourg Papers, II:130; Baudier, *Catholic Church in Louisiana*, 304-305.

149. Wilson and Lemann, *Lower Garden District*, 27; Bringier Papers, IV:106.

150. Family tradition maintained as late as 1921 that it was this engine and company who were involved in one of New Orleans' most famous ghost stories—the Lalaurie Haunted House. According to tradition, the owners of the home in the French Quarter, Dr. Louis Lalaurie and his wife Madame Delphine Macarty de Lopez Blanque Lalaurie subjected their slaves to torture and starvation. Unknown to their neighbors, the cruel treatment of their servants was only discovered upon the arrival of firemen to fight a fire started by the terrified slaves. Bringier tradition maintains that it was the engine housed at their Melpomene estate that arrived to fight the fire and that it was members

of the engine's company who discovered the tortured slaves. However, it is unlikely that the Bringier engine could have been involved in the famous event. The Lalaurie atrocities occurred in the early 1830s, several years before the Bringiers obtained possession of their Melpomene estate in 1838. Doradou's daughter Louise (Mrs. Martin Gordon, Jr.) claimed to have had dinner at the Lalaurie house on Royal Street shortly before the discovery of the mistreated slaves and observed how the slaves of the house reacted in terror for minor infractions such as dropping a tumbler. Gordon also maintained that when the fire occurred shortly after the dinner, the slave that dropped the tumbler was found chained to a post and positioned so that food trays were just out of the unfortunate slave's reach. Bringier Papers, I:3; Bringier Papers, II:187; Bringier Papers, IV:90, 147-147a; Mitchell, *Classic New Orleans*, 89.

151. The Gordons' financial problems were so serious that they were forced to sell their home on Royal Street to Judge Alonzo Morphy. The judge's son Paul would become the internationally renown chess champion. Today the building is used as Brennan's restaurant. Bringier Papers, II:178-179; Bringier Papers, IV:92.

152. Bringier Papers, II:93, 179, 195-196; Bringier Papers, IV:71, 92; Wilson and Lemann, *Lower Garden District*, 27.

153. Bringier Papers, IV:94-94a, 95a, 103.

154. Quoted by Woods in Bringier Papers, II:197-198.

155. Bringier Papers, IV:101; Parrish, *Richard Taylor*, 42.

156. Wilhelminia was the daughter of Hore Browse Trist. Included among the children of Wilhelminia and Colonel Wood was young Trist Wood who would spend his life collecting material and remembrances about his family. Bringier Papers, II:111.

157. Duncan and Nanine Kenner acquired ownership of Melpomene late in Aglae's life when she was in danger of losing the property and its furnishings because of financial problems. Without Aglae's knowledge, the Kenners paid her debts and saved the structure and the old granddame the embarrassment of foreclosure.

158. Bringier Papers, II:110; Bringier Papers, IV:94-94a, 101, 103-105; photocopy of *Illustrated American*, September 26, 1891 in Bringier Papers, IV:103.

Chapter Six

159. For a detailed description of the sugar production process on a plantation see Sitterson, *Sugar Country*, 112-156 and Walter Prichard, "Routine on a Louisiana Sugar Plantation under the Slavery Regime," *Mississippi Valley Historical Review* 14 (September 1927): 168-178; for a shorter, yet noteworthy, description see Glenn R. Conrad and Ray F. Lucas, *White Gold: A Brief History of the Louisiana Sugar Industry 1795-1995* (Lafayette, La.: Center for Louisiana Studies, 1995), 12-28.

160. Conrad and Lucas, *White Gold*, 17; John C. Rodrigue, *Reconstruction in the Cane Field: From Slavery to Free Labor in Louisiana's Sugar Parishes, 1862-1880* (Baton Rouge: Louisiana State University), 13; Sitterson, *Sugar Country*, 117-118; Ashland Plantation Record Book, January 20, 1852, Louisiana and Lower Mississippi Valley Collections, Hill Memorial Library, Louisiana State University, Baton Rouge, Louisiana (hereafter cited as Ashland Plantation Book).

161. Ashland Plantation Record Book, January 13 and 19, 1852; Conrad and Lucas, *White Gold*, 15.

162. Sitterson, *Sugar Country*, 113; Conrad and Lucas, *White Gold*, 14-15; Rodrigue, *Reconstruction in the Cane Field*, 13.

163. Joseph Karl Menn, *The Large Slaveholders of Louisiana–1860* (New Orleans: Pelican Publishing Co., 1964), 8, 63.

164. Rodrigue, *Reconstruction in the Cane Field*, 14.

165. Amedee Bringier to Stella, December 27, 1855, Louis A. Bringier Papers.

166. At the time the referenced measurement was made (1836-1839), Kenner was in the early stages of establishing Ashland as a grand plantation. Archeological examination of the Ashland sugarhouse shows that over the next twenty years, Kenner enlarged and modified the structure multiple times to increase its capacity. Ibid.; Bauer, *Leader Among Peers*, 46-47; Bringier Papers, IV:113; Jill-Karen Yakubik and Rosalinda Mendez, *Beyond the Great House: Archaeology at Ashland-Belle Helene Plantation*, (Baton Rouge: Louisiana Department of Culture, Recreation and Tourism–Division of Archaeology, n.d.), 12; note from Robert Judice to author, October 10, 2007.

167. Livre Contenans les affaires De M. Doradou Bringier, Louis A. Bringier Papers.

168. Duncan Farrar Kenner to William J. Minor, January 22, 1846, Duncan Farrar Kenner Papers, Louisiana and Lower Mississippi Valley Collections, Hill Memorial Library, Louisiana State University, Baton Rouge, Louisiana (hereafter cited as Kenner Papers); Conrad and Lucas, *White Gold*, 19; Bauer, *Leader Among Peers*, 47-48; the cost of $7,500 was for the sugar-making equipment alone and did not include the substantial costs needed for the building, tools, animals, etc., which were also needed for production of sugar and increased the cost to the totals mentioned above. Not all sugar-producing estates could afford some of the latest and elaborate equipment noted and instead produced their sugar by more primitive means. The 1860 census figures for Ascension Parish reveal that when all of the sugar estates in the parish with fifty or more slaves are considered, they averaged implement and machinery inventories totaling only $26,772 and ranged from $200,000 for the parish's largest estate which was owned by John Burnside to only $250 worth of equipment used by James Ventress on his property. See Menn, *Large Slaveholders of Louisiana*, 24, 122-124.

169. The Ascension Parish planter in 1860 with the largest investment in farm and sugar-making equipment was John Burnside of Houmas House plantation. A native of Ireland and a successful New Orleans merchant, Burnside paid a reported $1,000,000 for his property of nearly 24,000 acres. Housing 752 slaves at his Houmas House estate, Burnside reported to census takers that the equipment and machinery on the estate was valued at $200,000. See Scarborough, *Masters of the Big House*, 126-127; Menn, *Large Slaveholders of Louisiana*, 121-124; Succession of Duncan Farrar Kenner in Kenner Papers.

170. In more modern times, diesel and natural gas engines have been used to power the mills. Conrad and Lucas, *White Gold*, 19-20.

171. Ibid., 20.

172. The definition of a hogshead differed some from source to source. As noted, the weight of a hogshead changed as the molasses content drained. In his annual publications in which he tracked the yearly sugar production figures in Louisiana, Pierre Champomier estimated that the average hogshead weighed 1,150 pounds; whereas, census returns used 1,000 pounds as the approximate weight of the large sugar containers. See Menn, *Large Slaveholders of Louisiana*, 112-113.

173. Ibid., 20-22; Sitterson, *Sugar Country*, 65-65; M. D. Bringier to his son M. S. Bringier, November 24, 1846, photocopy provided to the author by Robert Judice.

174. Conrad and Lucas, *White Gold*, 22-23; Bauer, *Leader Among Peers*, 67; Bringier Papers, II:198-198a; Bringier Papers, IV:87.

175. Bauer, *Leader Among Peers*, 252-255.

176. Ibid.; Susan E. Hocker, "Index to Early Louisiana Patents 1810-1890," (Miami, Ohio: Miami University Libraries, 2002) http://staff.lib.muohio.edu/~shocker/LAPAT/invent.php?iname=Bringier (accessed March 27, 2007). See Appendix B for complete list of Bringier Patents.

177. The honor of the largest single sugar crop belongs to the Bringier's neighbor John Burnside who owned an estate of 7,600 improved acres and produced a crop in 1861 of 7,652 hogsheads. Scarborough, *Masters of the Big House*, 136-137.

178. Pierre A. Champomier, *Statement of the Sugar Crop Made in Louisiana, 1844-1861* (New Orleans: Cook, Young, and Company, 1845-1862); Menn, *Large Slaveholders of Louisiana*, 79, 121-122.

179. Bringier Papers, IV:138-139.

180. T. H. Kennedy to Allan Thomas, May 10, 1858, Bringier Papers, IV:139, 142-144.

181. With Louisiana law defining slaves as real property, M. S. Bringier's real property was valued at $250,000 and his personal property totaled $250,000. Amedee's real property was valued at $80,000 and his real property was estimated at $90,000. Menn, *Large Slaveholders of Louisiana*, 79, 121-122; Champomier, *Statement of the Sugar Crop*; Notarized statement of

partial property transfer by Louise Elizabeth Aglae Bringier to Louis Amedee Bringier, June 11, 1858, Louis A. Bringier Papers.

182. During the antebellum period, the Bringiers also intermittently produced a crop on their White Hall property in St. James Parish. During the ten-year period, two crops were produced on the land, including 153 hogsheads in 1851-1852 and seventy-five hogsheads in 1861-1862.

183. Champomier, *Statement of the Sugar Crop.* See Appendix C for listing of sugar crop totals for other members of the Bringier family.

184. Champomier, *Statement of the Sugar Crop*; Scarborough, *Masters of the Big House*, 143-145.

185. Even today when there continues to be a demand for Louisiana's unique tobacco, it can be produced only in the same small area it was during Doradou's time.

186. Bringier Papers, III:61, 134a; R. Reese Fuller, "Perique the Native Crop," *Louisiana Life* 23, no. 1 (Spring 2003): 46-48.

187. The Hermitage consisted of 690 improved and 140 unimproved acres. The 44,000 unimproved acres of the Houmas property made it the largest single property holding in Ascension Parish. As classified by historian Joseph Menn in his study of the slaveholders in Louisiana in 1860, a "large slaveholding" plantation is defined as an estate where fifty or more slaves were held.

188. Menn, *Large Slaveholders of Louisiana*, 121-122.

Chapter Seven

189. William J. Cooper and Thomas E. Terrill, *The American South: A History*, 2nd ed. (New York: The McGraw-Hill Companies, Inc., 1996), 199.

190. Roudenez was a specialist in setting sugar kettles and sold his services to a large number of sugar planters in the region.

191. Of the artisans and workers listed, Boudreaux, Lea, Malone, Garbiel, and Lardner were noted by Landry as being white and Roudenez as a free person of color. The others had no racial identification listed. James D. Wilson, Jr., "'To Live Together in Peace:' The Extraordinary Life and Times of Pierre Caliste Landry," (unpublished manuscript), 25-26.

192. Scarborough, *Masters of the Big House*, 176.

193. In his detailed analysis of the planter elite, *Masters of the Big House*, Scarborough specifically addresses the belief presented by some modern historians that slavery was such a repugnant institution that "southern planters *must* have felt guilty about owning slaves," by noting that there is little evidence in their writings to support the generalization of mass guilt among white Southerners. This was the case with the Bringiers. For in the thousands of family documents that survive, there is no indication of feelings of guilt or remorse towards slavery on their part. Ibid., 176-177, 417-421; Peter Kolchin, *American Slavery: 1619-1877* (New York: Hill and Wang, 1993), 191-193.

194. Scarborough, *Masters of the Big House*, 154-155.

195. Cooper and Terrill, *American South*, 199-202.

196. William Kauffman Scarborough, *The Overseer: Plantation Management in the Old South* (Baton Rouge: Louisiana State University Press, 1966), 5, 43, and 60.

197. Rodrigue, *Reconstruction in the Cane Field*, 18; Scarborough, *Masters of the Big House*, 164-165; Cooper and Terrill, *American South*, 202-203, 210; Bauer, *Leader Among Peers*, 52.

198. Sitterson, *Sugar Country*, 99; Wilson, "To Live Together in Peace," 26.

199. Quoted by Wilson, "To Live Together in Peace," 24.

200. Ibid., 23-26.

201. The unique career of the slave Pierre Caliste Landry did not end with the closing of "Joe and Caliste." After the store closed, probably because of wartime difficulties in obtaining items to barter and sell, Caliste went on to serve apprenticeships with the plantation's engineer and machinist and the estate's carpenter. He remained with the Bringiers at Houmas even after being emancipated. Leaving the plantation in 1866, he moved to Donaldsonville where he became involved in Reconstruction politics. In 1868 he was elected mayor, thus making him among the first African Americans in the country to hold such a position in a municipality. He later moved to New Orleans and served nearly fifty years as a clergyman in the Methodist Episcopal church. Ibid., 9, 26-27.

202. Quoted in Scarborough, *Masters of the Big House*, 206.

203. The practice of providing slaves small plots of land to cultivate on their own account was actually protected by Louisiana law. However, the state's slave code also noted that "All that a slave possesses, belongs to his master. . . ." Quoted by Richard Follett, *The Sugar Masters: Planters and Slaves in Louisiana's Cane World, 1820-1860* (Baton Rouge: Louisiana State University Press, 2005), 197.

204. Follett, *Sugar Masters*, 206; Scarborough, *Masters of the Big House*, 97-98; John W. Blassingame, ed., *Slave Testimony: Two Centuries of Letters, Speeches, Interviews and Autobiographies* (Baton Rouge: Louisiana State University, 1977), 392-393.

205. Bringier Notes, II:64.

206. Kolchin, *American Slavery*, 113; Yakubik and Mendez, *Beyond the Great House*, 22.

207. Kolchin, *American Slavery*, 114; Ledger of M. D. Bringier account by Martin Gordon, Jr., Louis A. Bringier Papers; Martin Gordon, Jr., to Benjamin Tureaud, December 30, 1852, Benjamin Tureaud Papers, Louisiana and Lower Mississippi Valley Collections, Hill Memorial Library, Louisiana State University, Baton Rouge; Bauer, *Leader Among Peers*, 54; Eugene Genovese, *Roll, Jordon, Roll: The World the Slaves Made* (New York: Pantheon Books, 1974), 551-552.

208. Genovese, *Roll, Jordon, Roll*, 524.

209. Livre Contenans les affaires De M. Doradou Bringier, Louis A. Bringier Papers; Genovese, *Roll, Jordon, Roll*, 524; Sitterson, *Sugar Country*, 61; Yakubik and Mendez, *Beyond the Great House*, 18; Jessie Poesch and Barbara SoRelle Bacot, eds., *Louisiana Buildings 1720-1940: The Historic American Buildings Survey* (Baton Rouge: Louisiana State University Press, 1997), 129-130; Menn, *Large Slaveholders of Louisiana*, 122.

210. This was approximately the same average for the number of persons per free household of the time. The 1860 census showed that there were 5.2 slaves per house on large plantations and 5.3 individuals in free households. Robert William Fogel, and Stanley L. Engerman, *Time on the Cross: The Economics of American Negro Slavery* (Boston: Little, Brown and Co., 1974), 155.

211. Menn, *Large Slaveholders of Louisiana*, 122.

212. Hermitage Plantation Journal, 1833, Louis A. Bringier Papers; DuBourg Papers, II:211, 234.

213. Sitterson, *Sugar Country*, 61; Joe Gray Taylor, *Negro Slavery in Louisiana* (Baton Rouge: Louisiana Historical Association, 1963), 51-52; M. D. Bringier Ledger, May 19, 1844, Louis A. Bringier Papers.

214. Actually the female slave Francis, age 40, was given a value of $3,100. However, her worth included the value of her six children—ages 5 to 15 years.

215. Mrs. M. D. Bringier, Inventory of Hermitage Plantation, January 23, 1858, Louis A. Bringier Papers; Act of Sale between Aglae Bringier and L. A. Bringier, June 11, 1858, Louis A. Bringier Papers.

216. Although they suffered a 12 percent higher mortality rate than white Americans, according to historians Robert Fogel and Stanley Engerman, the life expectancy of slaves in America was nearly the same as that of the developed countries of Europe such as France and Holland. Slaves actually enjoyed a longer lifetime expectancy than the free urban industrial workers in both the United States and Europe. Fogel and Engerman, *Time on the Cross*, 126.

217. Genovese, *Roll, Jordon, Roll*, 520; Taylor, *Negro Slavery in Louisiana*, 82; Fogel and Engerman, *Time on the Cross*, 154; Mrs. M. D. Bringier, Inventory of Hermitage Plantation, January 23, 1858, Louis A. Bringier Papers; see Appendix F for age and gender ratios of the Hermitage's slave force in 1858.

218. John W. Blassingame, *The Slave Community: Plantation Life in the Antebellum South* (New York: Oxford University Press, 1972), 78; Mrs. M. D. Bringier, Inventory of Hermitage Plantation, January 23, 1858, Louis A. Bringier Papers.

219. Kolchin, *American Slavery*, 4; Mrs. M. D. Bringier, Inventory of Hermitage Plantation, January 23, 1858, Louis A. Bringier Papers.

220. Mrs. M. D. Bringier, Inventory of Hermitage Plantation, January 23, 1858, Louis A. Bringier Papers; Kolchin, *American Slavery*, 114; Wilson, "To Live Together in Peace," 25; Fogel and Engerman, *Time on the Cross*, 117, 120; Martin Gordon, Jr., to Marius Ste. Colomb Bringier, August 5, 1848, Robert Judice Collection.

221. Kolchin, *American Slavery*, 139; Taylor, *Negro Slavery in Louisiana*, 123.

222. L. A. (Amedee) Bringier to Stella, January 25, 1856, Louis A. Bringier Papers; Walter Johnson, *Soul By Soul: Life Inside the Antebellum Slave Market* (Cambridge, Mass.: Harvard University Press, 1999), 122-123.

223. Kolchin, *American Slavery*, 139; Taylor, *Negro Slavery in Louisiana*, 124; Mrs. M. D. Bringier, Inventory of Hermitage Plantation, January 23, 1858, Louis A. Bringier Papers.

224. Amedee Bringier to Stella, January 25, 1856, Louis A. Bringier Papers; Scarborough, *Masters of the Big House*, 176.

225. Fogel and Engerman, *Time on the Cross*, 144; Bauer, *Leader Among Peers*, 57.

226. Quoted by Taylor, *Negro Slavery in Louisiana*, 181.

227. It is not known what consequences Old Pleasant faced for his damage. Quoted in Follett, *Sugar Masters*, 146.

228. DuBourg Papers, II:211, 234; Scarborough, *Masters of the Big House*, 428, 435, 442 and 463; Scarborough, *The Overseer*, 11.

229. Scarborough, *Masters of the Big House*, 166.

Chapter Eight

230. N. P. Banks to The President of the United States, September 11, 1863, in Fred Harvey Harrington, "A Peace Mission of 1863," *The American Historical Review* 46 (October 1940): 78, 82-83; Bauer, *Leader Among Peers*, 203; *The War of the Rebellion: A Compilation of the Official Records of the Union and Confederate Armies* (hereafter cited as *Official Records*), Series 3, Vol. 2 (Washington, D.C.: United States Government Printing Office, 1880-1901), 796; Bringier Papers, II:198a; Parrish, *Richard Taylor*, 326.

231. E. Kirby Smith to Maj. Gen. R. Taylor, January 24, 1864, *Official Records*, Series I, Vol. 34, Part 2, 911.

232. E. Kirby Smith to Maj. Gen. R. Taylor, March 13, 1864, *Official Records*, Series I, Vol. 34, Part 1, 493-494.

233. Joseph G. Tregle, Jr., "Thomas J. Durant, Utopian Socialism, and the Failure of Presidential Reconstruction in Louisiana," *Journal of Southern History* 45 (November 1979): 497.

234. The Bringier's home parish of Ascension was among the strongest unionist areas of the state. In the 1860 presidential election, Ascension was one of the few parishes to vote in favor of Stephen Douglas, the leading unionist among the various major Democratic candidates running that year. Even more demonstrative of the region's anti-secession feelings was the parish's vote for representatives to Louisiana's Secession Convention in 1861 when 63 percent of the votes cast in the parish went to candidates who opposed immediate secession. Parrish, *Richard Taylor*, 100-103, Bauer, *Leader Among Peers*, 176.

235. Quoted in Parrish, *Richard Taylor*, 105.

236. *Daily Delta*, January 2, 1861.

237. Ironically, Burnside, a wealthy native of Northern Ireland who owned the vast 12,000 acre Houmas House estate a short distance down river from the Hermitage, secretly was a Unionist at heart who did not reveal his true feelings until after the region was under Federal control. Scarborough, *Masters of the Big House*, 126-127, 343; Bauer, *Leader Among Peers*, 190-191; Russell, *My Diary, North and South*, 279.

238. Bauer, *Leader Among Peers*, 174-185.

239. Craig A. Bauer, "The Last Effort: The Secret Mission of the Confederate Diplomat, Duncan F. Kenner," *Louisiana History* 22, no. 1 (Winter 1981): 67-78.

240. Ibid., 79-94.

241. Richard Taylor, *Destruction and Reconstruction: Personal Experiences of the Late War* (D. Appleton and Company, 1879; reprint, edited by Charles P. Roland. Waltham, Mass.: Blaisdell Publishing Company, 1968) ix; Conrad, *Dictionary of Louisiana Biography*, s.v. "Richard Taylor," by Arthur W. Bergeron, Jr.

242. Taylor's sister Sarah was the Confederate president's first wife until her death only three months after they were married.

243. Taylor, *Destruction and Reconstruction*, 187.

244. Christopher G. Peña, *Touched By Fire: Battles Fought in the Lafourche District*. (Thibodaux, La.: C.B.P. Press, 1998), 111; Bergeron, "Richard Taylor."

245. Bringier Papers, IV:11-12a; Conrad, *Dictionary of Louisiana Biography*, s.v. "Allen Thomas," by Arthur W. Bergeron, Jr.

246. Bringier Papers, IV:22-28; "Mother in the Civil War: The War Time Reminiscences of Wilhelmine Trist," Trist Wood Papers #800, Southern Historical Collection, The Wilson Library, University of North Carolina Library, Chapel Hill, North Carolina, photocopy provided author by Robert Judice.

247. Bringier Papers, III:47.

248. When Amedee joined Scott's cavalry, he signed as an "independent." Independents were usually gentlemen enlistees who were granted certain privileges such as not being required to perform certain unpleasant soldierly duties such as guard and picket duty. As an independent, he took part in any engagements or extraordinary services in which the unit participated. Independents did not take an oath to the unit and thus they reserved the right to leave whenever they wanted. Naturally, because of the privileges they enjoyed, regular enlistees sometimes resented the independents and called them by the derisive term "peacocks." Ibid., 49, 53-54a.

249. Ibid., 48-49.

250. J. Scott to Friend (Amedee) Bringier, April 14, 1864, Louis A. Bringier Papers.

251. Record of L. Amedee Bringier by A. B. Booth, Commissioner, Louisiana Military Records, July 26, 1916, photocopy provided to the author by

Robert Judice. After the war Joseph Lancaster Brent married Rosella Kenner, the daughter of Duncan and Nanine Kenner.

252. Bringier Papers III:56-56a.

253. Ibid., 57; Bringier Papers, IV:26-26a.

254. L. A. Bringier to un-named officer, undated, Louis A. Bringier Papers; Arthur W. Bergeron, Jr., *Guide to Louisiana Confederate Military Units 1861-1865* (Baton Rouge: Louisiana State University Press, 1989), 50-51; Col. L. A. Bringier to Brigadier General J. L. Brent, March 18, 1865, photocopy provided to the author by Robert Judice.

255. Bringier Papers, III:51.

256. Ibid., 49; Bauer, *Leader Among Peers*, 193.

257. "Mother in the Civil War;" Bauer, *Leader Among Peers*, 193.

258. Bauer, *Leader Among Peers*, 193; Bringier Papers, II:101.

259. Sniping at Union ships on the river by Confederates along the banks continued throughout much of the war. In 1863, a rebel battery even set up near the site of the Bringier's White Hall property to harass Yankee vessels rounding the bend in the river near the plantation. Sternberg, *Along the River Road*, 173.

260. *Official Records*, Series I, Vol. 15, 796.

261. Bringier Papers, III:55-55a, 135-135a.

262. "Stel" (Stella Bringier) to "Mede" (Amedee Bringier), July 28, 1862, Louis A. Bringier Papers.

263. Family records claim that the house where they stayed was the deserted home of J. Madison Wells, who would later serve as reconstruction governor of Louisiana, and that included among the individuals who stayed with the family in the Moundville house was Bethiah Moore, the wife of Gov. Thomas O' Moore. Bringier Papers, II:100.

264. Bauer, *Leader Among Peers*, 199-200; Bringier Papers, II:100, 156; Bringier Papers, III:85.

265. B. Tureaud to Bringier, December 30, 1863, note in "Mother in the Civil War;" Bringier Papers, II:101-102.

266. Her trip to meet her husband was nearly ended in Brookhaven, Mississippi upon the surprise arrival of Federal troops under Col. Benjamin H. Grierson. Fortunately, as Grierson was on the last leg of his daring raid across Mississippi and Louisiana and was being sought by large numbers of Confederate military units, the Yankee soldiers did not stay long enough to prevent Octavie's reunion with her husband; Bringier Papers, III:88-89, 90a, 150.

267. Ibid., 89-91.

268. Ibid.

269. Charles Roland, *Louisiana Sugar Plantations During the American Civil War* (Leiden, Netherlands: E. J. Brill, 1957), 50; Bringier Papers, III:55-55a, 135-138a.

270. "Mother in the Civil War;" Bringier Papers, III:55-55a, 135-138a.

271. Quoted by Scarborough, *Masters of the Big House*, 355-356.
272. Ibid.; Bringier Papers, III:55-55a, 61, 134a-138a.
273. Joe Gray Taylor, *Louisiana Reconstructed: 1863-1877* (Baton Rouge: Louisiana State University Press, 1974), 14, 35-36; Bringier Papers, III:55.
274. Bringier Papers, III:55-55a.
275. J. N. B. (James Baxter) to James, November 1, 1864, copy in Hermitage Foundation Papers; Sitterson, *Sugar Country*, 218; Rodrigue, *Reconstruction in the Cane Fields*, 39-46.
276. James Baxter to James, May 9, 1865, copy in Hermitage Foundation Papers; Sitterson, *Sugar Country*, 216-217; Conrad and Lucas, *White Gold*, 34.
277. It is likely Baxter was either exaggerating his crop acreage to impress his financial backers or was citing figures from all of his properties in the region and not just the Hermitage, because the acreage noted was greater than that available on the Bringier property.
278. J. N. B. (James Baxter) to James, June 4, 1864, Hermitage Foundation Papers; Sitterson, *Sugar Country*, 216-217; Ludwell H. Johnson, "The Red River Campaign," in *The Civil War Battlefield Guide*, ed. Frances H. Kennedy, (Boston: Houghton Mifflin Company, 1990), 163-164.
279. Roland, *Louisiana Sugar Plantations*, 71-72; *Official Records*, Series I, Vol. 41, 991, Vol. 48, 167-168, 803, 976.

Chapter Nine

280. John D. Winters, *The Civil War in Louisiana* (Baton Rouge: Louisiana State University, 1963), 428-429; Rodrigue, *Reconstruction in the Cane Fields*, 59; Sitterson, *Sugar Country*, 226.
281. John B. Boles, *The South Through Time: A History of an American Region* (Englewood Cliffs, NJ: Prentice Hall, 1995), 350-351.
282. Eric Foner, *Reconstruction: America's Unfinished Revolution, 1863-1877* (New York: Harper and Row Publishers, 1988), 183-184.
283. Ibid., 190-191; Rodrigue, *Reconstruction in the Cane Fields*, 62.
284. Bringier Papers, III:62-62a, 138a.
285. Although Stella and Amedee gained possession of the Hermitage, actual ownership remained shared with his mother Aglae. Later her share was transferred to the youngest of her sons, Doradou, who later disposed of his share of the estate to his brother-in-law Duncan Kenner. Bringier Papers, III:144.
286. Quoted by Scarborough, *Masters of the Big House*, 380, 381; Sitterson, *Sugar Country*, 231-232.
287. Bringier Papers, III:58aa-60; Bauer, *Leader Among Peers*, 202-203.
288. Bringier Papers, III:59-60; Bringier Papers, IV:42, 73.
289. Sitterson, *Sugar Country*, 231-232.
290. Ibid., 301-302; Roland, *Louisiana Sugar Plantations During the American Civil War*, 139.
291. Bringier Papers, II:147a-148, Bringier Papers, IV:113; Notarized state-

ment of ownership of Houmas and Hermitage Plantations by Nicholas Browse Trist, January 13, 1870, Louis A. Bringier Papers; Richard Follett, *Documenting Louisiana Sugar 1845-1917* (Sussex, England: University of Sussex, 2008) [database on-line].

292. Notarized statement of sale of Hermitage Plantation, February 12, 1869, Louis A. Bringier Papers; Bauer, *Leader Among Peers*, 245-246.

293. See Appendix B; United States Patent Office, Patent 409, 847- Decorticator for Ramie, August 27, 1889; Bauer, *Leader Among Peers*, 252; Sitterson, *Sugar Country*, 276.

294. Boles, *South Through Time*, 350-351; Bringier Papers, III:142.

295. Stella Bringier to Brig. Gen. J. L. Brent, January 30, 1865, photocopy provided to the author by Robert Judice.

296. Bringier Papers, III:63a-64, 144-145.

297. Ibid., 144-145.

298. Ibid.

299. Louis Amedee Bringier to Stella, December 27, 1855, Louis A. Bringier Papers.

300. Louis Amedee Bringier to Stella, January 6, 1858, Louis A. Bringier Papers.

301. Scarborough, *Masters of the Big House*, 49-50.

302. Ibid., 49-50; notarized statement of sale of the Hermitage Plantation, February 12, 1869, Louis A. Bringier Papers; notarized statement of ownership of Houmas and Hermitage Plantations by Nicholas Browse Trist, January 13, 1870, Louis A. Bringier Papers.

303. Bringier Papers, III:144-145; Bauer, *Leader Among Peers*, 247; Sitterson, *Sugar Country*, 362-363.

304. Quoted in Sitterson, *Sugar Country*, 362; Bringier Papers, III:139, 145a, 158, 160.

305. Bauer, *Leader Among Peers*, 243, 248, 284, 289.

306. Ibid., 251-255, 289.

307. Craig Bauer and Mark LaFlaur, *Dictionary of American National Biography* (New York: Oxford Press, 2002), s.v. "Duncan F. Kenner."

308. Bringier Papers, III:108-111; Bringier Papers, IV:2-4.

309. Bringier Papers, IV:1, 13.

310. Ibid., 5-15; Horace Samuel Merrill, *Bourbon Leader: Grover Cleveland and the Democratic Party* (Boston: Little, Brown and Company, 1957), 202.

311. The large loans between Aglae and the Taylors were made without the knowledge of most of the other family members. Only her son-in-law Martin Gordon knew of the notes. Of all the members of Bringier family, Aglae depended most on Gordon for financial advice. As one of the leading sugar factors in New Orleans, Gordon had a great deal of confidence in the future of the sugar industry in the state and thus advised Aglae that her investment in the Taylor's Fashion estate was not only a opportunity to provide help to her

daughter but also a sound financial investment. Bringier Papers, IV:40-41a; Parrish, *Richard Taylor*, 456-457.

312. Parrish, *Richard Taylor*, 457-458; Bauer, *Leader Among Peers*, 249-250.
313. Parrish, *Richard Taylor*, 459-460, 477-496.
314. Ibid., 447-455; Bringier Papers, IV:29-29a; Aglae Bringier to Marius Ste. Colomb Bringier, July 15, 1865, Robert Judice Collection.
315. Bringier Papers, IV:30.
316. Ibid., 31-33.
317. Bringier Papers, IV:34, 36-37.
318. Ibid., 96.
319. Bringier Papers, II:102.
320. Ibid., 102a.
321. Christovich and Evans, *Esplanade Ridge*, 98, 100.
322. Note dated November 23, 1865, Bringier Papers, IV:96.
323. Bringier Papers, IV:96a-97a.
324. Benjamin and Aglae Trudeau's (daughter of Doradou and Aglae Bringier) Tezcuco plantation house actually remained in the hands of Bringier family members until 1946. Bringier Papers, II:103, 110-111; Bringier Papers, IV:71-72; Copy of partnership agreement between Aglae Bringier and L. A. Bringier, June 16, 1858, Orleans Parish Archives; Sternberg, *Along the River Road*, 176.

Chapter Ten

325. William Ivy Hair, *Bourbonism and Agrarian Protest: Louisiana Politics 1877-1900* (Baton Rouge: Louisiana State University Press, 1969), 37.
326. Ibid., 37-38; *Daily Picayune*, September 23, 1877.
327. Ibid.
328. Follett, *Documenting Louisiana Sugar*.
329. Bringier Papers, IV:75-78a; "New Hermitage" brochure and other miscellaneous documents and notes on the New Hermitage real estate development and St. Elmo community in the possession of Robert Judice.
330. "New Hermitage" brochure.
331. At the time of the Judice purchase, there were also two twentieth-century structures located about twenty-five feet from the house. One was a chain-link fence to keep roaming cattle away and the other was a small garage.
332. Undated note from Robert Judice to the author; Sexton, *Vestiges of Grandeur*, 82; Sternberg, *Along the River Road*, 181-182.
333. Undated note from Robert Judice to the author.
334. Ibid.; Poesch, "Furniture of the River Road Plantations," 1190, 1192.

Appendix A

Burials in the Bringier Tomb - Donaldsonville, Louisiana

Name	Date of Death
Bringier, Aglae DuBourg (Mrs. M. D.)	1878
Bringier, Augustine Tureaud (Mrs. M. S.)	March 17, 1887
Bringier, Felicie Augustine	June 7, 1915
Bringier, Louis Amadee	January 9, 1897
Bringier, Marius Pons	(Moved from St. James)
Bringier, Marius St. Colomb	August 22, 1884
Bringier, Marius St. Colomb, Jr.	October 5, 1905
Bringier, Martin Doradou	1887
Bringier, Michel Doradou	1847 (first to be buried in the tomb)
Bringier, Stella Tureaud (Mrs. L. A.)	November 11, 1911
Colomb, L. Arthur	May 6, 1903
Colomb, Louis and family	—
Gordon, Mary Wilhemine Trist	—
Kenner, Duncan Farrar	July 3, 1887
Kenner, George D.	January 30, 1883
Kenner, Nanine Bringier (Mrs. D. F.)	November 6, 1911
Simpson, Blanche Kenner	March 26, 1890
Thomas, Allen (General)	December 1907
Thomas, Allen, Jr.	January 20, 1918
Thomas, Anne Octavie Bringier (Mrs. Allen Thomas)	November 20, 1917
Trist, Hore Browse	November 16, 1856
Trist, Rosella Bringier (Mrs. H. B.)	July 20, 1849
Tureaud, A. D. and family	(Moved from St. James)

Name	Date of Death
Tureaud, Aglae Bringier (Mrs. B.)	—
Tureaud, Benjamin	1883
Tureaud, Benjamin, Jr.	August 30, 1880
Tureaud, Louise Bringier (Mrs. A.D.)	November 1865
Watermann, John R. (Brother-in-law L. Arthur Colomb)	May 16, 1876
Wilson, Melanie Bringier	1863

APPENDIX B

Bringier Family Patents

Inventor(s)	Title	Patent Number	Date
Louis A. Bringier & N. B. Trist	Cane scraper	92,420	1869
Louis A. Bringier	Back band for plow harness	217,983	1879
Louis A. Bringier	Tail board for wagons	252,996	1882
Louis A. Bringier	Cultivator	293,221	1884
Marius S. Bringier	Improved steam boiler	25,802	1859
Marius S. Bringier	Evaporating Pan	70,690	1867
Marius S. Bringier	Improved process for extracting saccharine matters from sugar-cane	70,691	1867
Marius S. Bringier	Mode of purifying water	81,979	1868
Marius S. Bringier	Food for domestic animals	87,821	1869
Marius S. Bringier	Cane juice evaporator	94,942	1869
Marius S. Bringier	Cane juice evaporator	96,081	1869

Appendix B

Inventor(s)	Title	Patent Number	Date
Marius S. Bringier	Steam generator	98,021	1869
Marius S. Bringier	Improvement in apparatus for extracting saccharine matter from sugarcane	124,030	1872
Marius S. Bringier	Improvement to apparatus for extracting saccharine matter from sugarcane & etc.	140,461	1873
Marius S. Bringier	Apparatus for extracting saccharine matter from sugarcane & etc.	141,316	1873
Marius S. Bringier	Apparatus for extracting saccharine liquor cane	179,679	1876

For additional information, including applications and drawings, of each patent see *Google Patents* on the internet.

APPENDIX C

Crop Production Figures and Ownership For Bringier Family Plantations 1844-1889*

Year	Parish	Planter	Plantation	Hogs-heads	# of 1,000 lbs	Notes
1844	Ascension	D. F. Kenner		1,156	1,200	
1844	Ascension	H. B. Trist		566	566	
1844	Ascension	M. D. Bringier		505	505	
1844	Ascension	Louis Colomb		500	520	
1844	Ascension	M. D. Bringier et Son		1,170	1,170	
1845	Ascension	D. F. Kenner		965	1,100	
1845	Ascension	H. B. Trist		388	460	
1845	Ascension	M. D. Bringier		660	690	
1845	Ascension	Louis Colomb		452	500	
1845	Ascension	D. S. Bringier & Co.		966	1,060	
1845	Ascension	D. S. Bringier & Co.		204	220	
1849-1850	Ascension	D. F. Kenner	Ashland		580	

177

Appendix C

Year	Parish	Planter	Plantation	Hogs-heads	# of 1,000 lbs	Notes
1849-1850	Ascension	H. B. Trist			735	"Partially lost by overflow."
1849-1850	Ascension	M. D. Bringier	Hermitage		380	"Partially lost by overflow."
1849-1850	Ascension	Louis Colomb			242	
1849-1850	Ascension	M. D. Bringier	Houmas		570	
1849-1850	Ascension	M. D. Bringier	"do back"		425	refers to Brulé property
1849-1850	St. James	Mrs. M. D. Bringier	White Hall		251	
1850-1851	Ascension	D. F. Kenner	Ashland		859	vacuum
1850-1851	Ascension	H. B. Trist	Bowden		632	Rillieux
1850-1851	Ascension	Mrs. M. D. Bringier	Hermitage		350	steam power
1850-1851	Ascension	Mrs. Louis Colomb			245	steam power
1850-1851	Ascension	Mrs. M. D. Bringier	Houmas		525	vacuum
1850-1851	Ascension	Mrs. M. D. Bringier	Bruslie		425	refers to Brulé property/ steam power
1850-1851	St. Charles	Richard Taylor	Fashion		175	steam power
1850-1851	St. James	Mrs. M. D. Bringier	White Hall		46	steam power

Crop Production Figures for 1844-1889

Year	Parish	Planter	Plantation	Hogs-heads	# of 1,000 lbs	Notes
1850-1851	St. James	Mrs. Tureaud & Son	Union		450	steam power
1851-1852	Ascension	D. F. Kenner	Ashland		710	vacuum
1851-1852	Ascension	H. B. Trist	Bowden		595	Rillieux
1851-1852	Ascension	Mrs. M. D. Bringier	Hermitage		496	steam power
1851-1852	Ascension	Mrs. Louis Colomb			230	"partly overflowed"
1851-1852	Ascension	Mrs. M. D. Bringier	Houmas		607	vacuum
1851-1852	Ascension	Mrs. M. D. Bringier	Brulé		0	1250 bbls syrup/steam power
1851-1852	St. Charles	Richard Taylor	Fashion		357	steam power
1851-1852	St. James	Mrs. M. D. Bringier	White Hall		298	steam power
1851-1852	St. James	Mrs. Tureaud & Son	Union		498	steam power
1852-1853	Ascension	D. F. Kenner	Ashland		1,169	vacuum
1852-1853	Ascension	H. B. Trist	Bowden		600	Rilleux
1852-1853	Ascension	Mrs. M. D. Bringier	Hermitage		560	steam power
1852-1853	Ascension	Mrs. Louis Colomb			502	steam power
1852-1853	Ascension	Mrs. M. D. Bringier	Houmas		1400	vacuum

Appendix C

Year	Parish	Planter	Plantation	Hogs-heads	# of 1,000 lbs	Notes
1852-1853	Ascension	Mrs. M. D. Bringier	Brulé		526	steam power
1852-1853	St. Charles	Richard Taylor	Fashion		530	steam power
1852-1853	St. James	Mrs. M. D. Bringier	White Hall		153	steam power
1852-1853	St. James	Mrs. Tureaud & Son	Union		610	steam power
1853-1854	Ascension	D. F. Kenner	Ashland		1,370	vacuum
1853-1854	Ascension	H. B. Trist	Bowden		810	Rillieux Apparatus
1853-1854	Ascension	Mrs. M. D. Bringier	Hermitage		662	steam power
1853-1854	Ascension	Mrs. Louis Colomb			710	steam power
1853-1854	Ascension	Mrs. M. D. Bringier	Houmas		2,400	2 sugar houses, 1 vacuum
1853-1854	Ascension	Mrs. M. D. Bringier	Brulé		0	steam power
1853-1854	St. Charles	Richard Taylor	Fashion		500	steam power
1853-1854	St. James	Mrs. M. D. Bringier	White Hall		0	steam power
1853-1854	St. James	Mrs. Tureaud	Union		850	steam power
1854-1855	Ascension	D. F. Kenner	Ashland		1,397	vacuum
1854-1855	Ascension	H. B. Trist	Bowden		755	Rillieux Apparatus
1854-1855	Ascension	Mrs. M. D. Bringier	Hermitage		530	steam power

Crop Production Figures for 1844-1889

Year	Parish	Planter	Plantation	Hogs-heads	# of 1,000 lbs	Notes
1854–1855	Ascension	Mrs. Louis Colomb			488	steam power
1854–1855	Ascension	Mrs. M. D. Bringier	Houmas		2,100	2 sugar houses, 1 vacuum
1854–1855	St. Charles	Richard Taylor	Fashion		360	steam power
1854–1855	St. James	Mrs. M. D. Bringier	White Hall		0	steam power
1854–1855	St. James	Mrs. Tureaud	Union		550	steam power
1855–1856	Ascension	D. F. Kenner	Ashland		570	vacuum
1855–1856	Ascension	H. B. Trist	Bowden		500	Rillieux Apparatus
1855–1856	Ascension	Mrs. M. D. Bringier	Hermitage		395	Open Steam Train
1855–1856	Ascension	Mrs. M. D. Bringier	Houmas		1,100	2 sugar houses, 1 vacuum
1855–1856	Ascension	Mrs. Louis Colomb			367	steam power
1855–1856	St. Charles	Richard Taylor	Fashion		644	Open Steam Train
1855–1856	St. James	Mrs. M. D. Bringier	White Hall		0	steam power
1855–1856	St. James	Mrs. Tureaud	Union		325	steam power
1856–1857	Ascension	D. F. Kenner	Ashland		342	vacuum
1856–1857	Ascension	H. B. Trist	Bowden		200	Rillieux Apparatus
1856–1857	Ascension	Mrs. M. D. Bringier	Hermitage		145	Open Steam Train

Appendix C

Year	Parish	Planter	Plantation	Hogs-heads	# of 1,000 lbs	Notes
1856-1857	Ascension	Mrs. M. D. Bringier	Houmas		675	2 sugar houses, 1 vacuum
1856-1857	St. Charles	Richard Taylor	Fashion		0	Open Steam Train
1856-1857	St. James	Mrs. M. D. Bringier	White Hall		0	steam power
1856-1857	St. James	Mrs. Tureaud	Union		300	steam power
1857-1858	Ascension	D. F. Kenner	Ashland		1,080	vacuum
1857-1858	Ascension	H. B. Trist	Bowden		550	Rillieux Apparatus
1857-1858	Ascension	Mrs. M. D. Bringier	Hermitage		900	Open Steam Train
1857-1858	Ascension	Mrs. Louis Colomb			517	steam power
1857-1858	Ascension	Mrs. M. D. Bringier	Houmas		1,140	2 sugar houses, 1 vacuum
1857-1858	St. Charles	Richard Taylor	Fashion		520	Open Steam Train
1857-1858	St. James	Mrs. M. D. Bringier	White Hall		0	steam power
1857-1858	St. James	Mrs. Tureaud	Union		430	steam power
1858-1859	Ascension	D. F. Kenner	Ashland		2002	vacuum
1858-1859	Ascension	L. A. Bringier	Hermitage		917	Open Steam Train
1858-1859	Ascension	Mrs. M. D. Bringier	Houmas		2,155	2 sugar houses, 1 vacuum
1858-1859	Ascension	Mrs. Louis Colomb			390	steam power

Crop Production Figures for 1844-1889

Year	Parish	Planter	Plantation	Hogs-heads	# of 1,000 lbs	Notes
1858-1859	St. Charles	Richard Taylor	Fashion		388	Open Steam Train
1858-1859	St. James	Mrs. M. D. Bringier	White Hall		0	steam power
1858-1859	St. James	Mrs. Tureaud	Union		560	steam power
1859-1860	Ascension	D. F. Kenner	Ashland		1,500	2 sugar houses, 1 vacuum, 1 Rillieux apparatus
1859-1860	Ascension	L. A. Bringier	Hermitage		280	Open Steam Train
1859-1860	Ascension	Mrs. M. D. Bringier	Houmas		1,350	2 sugar houses, 1 vacuum
1859-1860	Ascension	Mrs. Louis Colomb			220	steam power
1859-1860	St. Charles	Richard Taylor	Fashion		482	Open Steam Train
1859-1860	St. James	Mrs. M. D. Bringier	White Hall		0	steam power
1859-1860	St. James	Mrs. Tureaud	Union		425	steam power
1860-1861	Ascension	D. F. Kenner	Ashland		940	2 sugar houses, 1 vacuum, 1 Rillieux Apparatus
1860-1861	Ascension	L. A. Bringier	Hermitage		385	Open Steam Train
1860-1861	Ascension	Mrs. M. D. Bringier	Houmas		700	2 sugar houses, 1 vacuum
1860-1861	Ascension	Mrs. Louis Colomb			309	steam power

Appendix C

Year	Parish	Planter	Plantation	Hogs-heads	# of 1,000 lbs	Notes
1860-1861	St. Charles	Richard Taylor	Fashion		544	Open Steam Train
1860-1861	St. James	Mrs. M. D. Bringier	White Hall		0	steam power
1860-1861	St. James	Mrs. Tureaud	Union		398	steam power
1861-1862	Ascension	D. F. Kenner	Ashland		2,150	2 sugar houses, 1 vacuum, 1 Rillieux apparatus
1861-1862	Ascension	L. A. Bringier	Hermitage		1,250	Open Steam Train
1861-1862	Ascension	Mrs. Louis Colomb			450	steam power
1861-1862	Ascension	Mrs. M. D. Bringier	Houmas		2,000	2 sugar houses, 1 vacuum
1861-1862	St. James	Mrs. M. D. Bringier	White Hall		75	steam power
1861-1862	St. James	Mrs. Tureaud	Union		652	steam power
1861-1862	St. Charles	Richard Taylor	Fashion		930	Open Steam Train

Crop Production Figures for 1844-1889

Postwar Data

Year	Parish	Planter	Plantation	Sugar	Note
1868-1869	Ascension	D. F. Kenner	Ashland	0	no report for the year
1868-1869	Ascension	D. F. Kenner	Bowden	350	Rillieux
1868-1869	Ascension	Bringier & Co.	Hermitage	300	steam & vacuum
1868-1869	Ascension	Mrs. Louis Colomb	Bocage	68	sk & op (steam, kettles and open pan)
1868-1869	Ascension	Benj. Tureaud & Co.	Houmas	450	st & v (steam, vacuum and centrifugals)
1869-1870	Ascension	D. F. Kenner	Ashland	116	stv & c
1869-1870	Ascension	D. F. Kenner	Bowden	290	Rillieux
1869-1870	Ascension	Bringier & Co.	Hermitage	206	stv & c
1869-1870	Ascension	Benj. Tureaud & Co.	Houmas/Brulé	525	st & v
1870-1871	Ascension	D. F. Kenner	Ashland	352	stv & c
1870-1871	Ascension	D. F. Kenner	Bowden	348	Rillieux
1870-1871	Ascension	Bringier & Co.	Hermitage	465	stv & c
1870-1871	Ascension	Benj. Tureaud & Co.	Houmas/Brulé	0	steam power
1870-1871	Ascension	Benj. Tureaud & Co.	Houmas	809	stv & c

Appendix C

Year	Parish	Planter	Plantation	Sugar	Note
1871-1872	Ascension	D. F. Kenner	Ashland	363	stv & c
1871-1872	Ascension	D. F. Kenner	Bowden	285	Rillieux
1871-1872	Ascension	Bringier & Co.	Hermitage	442	stv & c
1871-1872	Ascension	Benj. Tureaud & Co.	Houmas/Brulé	0	steam power
1871-1872	Ascension	Benj. Tureaud & Co.	Houmas	685	stv & c
1872-1873	Ascension	D. F. Kenner	Ashland	194	stv & c
1872-1873	Ascension	D. F. Kenner	Bowden	242	Rillieux
1872-1873	Ascension	Bringier & Co.	Hermitage	183	stv & c
1872-1873	Ascension	Benj. Tureaud & Co.	Houmas/Brulé	210	steam power
1872-1873	Ascension	Benj. Tureaud & Co.	Houmas	505	stv & c
1873-1874	Ascension	D. F. Kenner	Ashland	296	stv & c
1873-1874	Ascension	D. F. Kenner	Bowden	334	Rillieux
1873-1874	Ascension	Bringier & Co.	Hermitage	350	stv & c
1873-1874	Ascension	Benj. Tureaud & Co.	Houmas/Brulé	150	steam power
1873-1874	Ascension	Benj. Tureaud & Co.	Houmas	300	stv & c
1874-1875	Ascension	D. F. Kenner	Ashland	424	stv & c
1874-1875	Ascension	D. F. Kenner	Bowden	384	Rillieux

Crop Production Figures for 1844-1889

Year	Parish	Planter	Plantation	Sugar	Note
1874-1875	Ascension	Bringier & Co.	Hermitage	400	stv & c
1874-1875	Ascension	Benj. Tureaud & Co.	Houmas/Brulé	0	steam power
1874-1875	Ascension	Benj. Tureaud & Co.	Houmas	527	stv & c
1875-1876	Ascension	D. F. Kenner	Ashland	415	stv & c
1875-1876	Ascension	D. F. Kenner	Bowden	555	Rillieux
1875-1876	Ascension	Bringier & Co.	Hermitage	330	stv & c
1875-1876	Ascension	Benj. Tureaud & Co.	Houmas/Brulé	0	steam power
1875-1876	Ascension	Benj. Tureaud & Co.	Houmas	730	stv & c
1876-1877	Ascension	D. F. Kenner	Ashland	481	stv & c
1876-1877	Ascension	D. F. Kenner	Bowden	475	Rillieux
1876-1877	Ascension	Bringier & Co.	Hermitage	420	stv & c
1876-1877	Ascension	Benj. Tureaud & Co.	Houmas/Brulé	0	steam power
1876-1877	Ascension	Benj. Tureaud & Co.	Houmas	630	stv & c
1877-1878	Ascension	D. F. Kenner	Ashland	273	stv & c
1877-1878	Ascension	D. F. Kenner	Bowden	274	Rillieux
1877-1878	Ascension	D. F. Kenner	84 miles fr/ NO	0	Land acquired from Mrs. J. Landry

Appendix C

Year	Parish	Planter	Plantation	Sugar	Note
1877-1878	Ascension	Bringier & Co.	Hermitage	105	stv & c
1877-1878	Ascension	Benj. Tureaud & Co.	Houmas/Brulé	0	steam power
1877-1878	Ascension	Benj. Tureaud & Co.	Houmas	370	stv & c
1878-1879	Ascension	D. F. Kenner	Ashland	335	stv & c
1878-1879	Ascension	D. F. Kenner	Bowden	520	Rillieux
1878-1879	Ascension	D. F. Kenner	84 miles fr/ NO	0	Land acquired from Mrs. J. Landry
1878-1879	Ascension	Bringier & Co.	Hermitage	408	stv & c
1878-1879	Ascension	Benj. Tureaud & Co.	Houmas/Brulé	0	steam power
1878-1879	Ascension	Benj. Tureaud & Co.	Houmas	760	stv & c
1879-1880	Ascension	D. F. Kenner	Ashland	355	stv & c
1879-1880	Ascension	D. F. Kenner	Bowden	450	Rillieux
1879-1880	Ascension	D. F. Kenner	84 miles fr/ NO	0	
1879-1880	Ascension	L. A. Bringier	Hermitage	0	stv & c
1879-1880	Ascension	Benj. Tureaud & Co.	Houmas/Brulé	0	steam power
1879-1880	Ascension	Benj. Tureaud & Co.	Houmas	880	stv & c

Crop Production Figures for 1844-1889

Year	Parish	Planter	Plantation	Sugar	Note
1880-1881	Ascension	D. F. Kenner & J. L. Brent	Ashland	938	stv & c (Kenner also produced 2,824 hogsheads of rice during the year.)
1880-1881	Ascension	D. F. Kenner & J. L. Brent	Bowden	450	Rillieux
1880-1881	Ascension	D. F. Kenner & L. A. Bringier	Hermitage	543	stv & c
1880-1881	Ascension	D. F. Kenner & Benjamin Tureaud	Houmas	672	stv & c
1881-1882	Ascension	D. F. Kenner & J. L. Brent	Ashland	530	stv & c
1881-1882	Ascension	D. F. Kenner & J. L. Brent	Bowden	0	Rillieux
1881-1882	Ascension	D.F. Kenner & L. A. Bringier	Hermitage	193	stv & c
1881-1882	Ascension	D. F. Kenner & Benjamin Tureaud	Houmas	484	stv & c
1882-1883	Ascension	D. F. Kenner & J. L. Brent	Ashland	0	stv & c

Appendix C

Year	Parish	Planter	Plantation	Sugar	Note
1882-1883	Ascension	D. F. Kenner & J. L. Brent	Bowden	579	Rillieux
1882-1883	Ascension	D. F. Kenner & L. A. Bringier	Hermitage	489	stv & c
1882-1883	Ascension	D. F. Kenner & Benjamin Tureaud	Houmas	1,079	stv & c
1883-1884	Ascension	D. F. Kenner & J. L. Brent	Ashland	0	stv & c
1883-1884	Ascension	D. F. Kenner & J. L. Brent	Bowden	1,053	Rillieux
1883-1884	Ascension	D. F. Kenner & L. A. Bringier	Hermitage	415	stv & c
1883-1884	Ascension	D. F. Kenner & Benjamin Tureaud	Houmas	891	stv & c
1884-1885	Ascension	D. F. Kenner & J. L. Brent	Ashland	0	stv & c
1884-1885	Ascension	Brent	Bowden	1,014	Rillieux
1884-1885	Ascension	D. F. Kenner & L. A. Bringier	Hermitage	209	stv & c

Crop Production Figures for 1844-1889

Year	Parish	Planter	Plantation	Sugar	Note
1884-1885	Ascension	D. F. Kenner & Benjamin Tureaud	Houmas	652	stv & c
1885-1886	Ascension	D. F. Kenner & J. L. Brent	Ashland	0	stv & c
1885-1886	Ascension	D. F. Kenner & J. L. Brent	Bowden	1,194	Rillieux
1885-1886	Ascension	D. F. Kenner	Hermitage	467	stv & c
1885-1886	Ascension	D. F. Kenner	Houmas	891	stv & c
1886-1887	Ascension	D. F. Kenner & J. L. Brent	Ashland	0	stv & c
1886-1887	Ascension	D. F. Kenner & J. L. Brent	Bowden	954	Rillieux
1886-1887	Ascension	D. F. Kenner	Hermitage	429	stv & c
1886-1887	Ascension	D. F. Kenner	Houmas	684	stv & c
1887-1888	Ascension	D. F. Kenner & J. L. Brent	Ashland	0	sprm (steam powered)
1887-1888	Ascension	D. F. Kenner & J. L. Brent	Bowden	1,170	Rillieux
1887-1888	Ascension	D. F. Kenner	Hermitage	329	stv & c
1887-1888	Ascension	D. F. Kenner	Houmas	1,040	stv & c
1888-1889	Ascension	D. F. Kenner & J. L. Brent	Ashland	0	sprm
1888-1889	Ascension	D. F. Kenner & J. L. Brent	Bowden	1,412	Rillieux

* See P.A. Champomier, *Statement of Sugar Crops Made in Louisiana* (1844-1862) and Alceé Bochereau, *Statement of Sugar and Rice Crops Made in Louisiana* (1879-1887).

Appendix D

*Hermitage Slave Inventory of 1858**

Slaves names	Age	Notes	# of Children	Value
Males				
Augustine	70			$1,000
Anthony	55	blind in one eye		$800
Alfred	27			$1,400
Aleck	33			$1,200
Antoine	50	sickly		$900
Bunell	32			$1,400
Baptiste	20			$1,400
Bastien	20			$1,400
Ben	32			$1,400
Ben Little	19			$1,300
Benjamin	19			0
Bob	65			$300
Christmas	40	blacksmith		$1,800
Simon	65			$200
Charles	28			$1,400
New Charles	29			$1,400
Currey	34			$1,000
Fed	34	(not on Eng. inven)		$1,300
Frederick	30			$1,300
Frank	40	consumptive		$200
George	25	engineer		$1,600
Henry	32			$800
Honoré	30			$1,300
Jesse	48	engineer		$1,400
New Jesse	40			$1,400
Janvier	30			$1,400
Jerry	36			$1,300
Joe	48			$800
Jefferson	50			$700

Hermitage Slave Inventory of 1858

Slaves names	Age	Notes	# of Children	Value
Jack	22			$1,400
Joseph	27			$1,400
John	58			$200
Jean Louis	40	affected with hernia		$700
Joe Congo	70			$200
Louis	55			$500
Martial	21			$1,400
Michel	50			$500
Ned	50			$400
Nathan	26			$1,400
Peter	50			$400
New Peter	34			$1,200
Peter Jones	44			$1,100
Perry	35			$1,200
Randall	27			$1,400
Robert	22			$1,300
Reuben	28			$1,300
Raphael	50			$400
Sam	38			$1,200
New Sam	38			$1,100
Simon	72	carpenter		$700
Little Simon	25			$1,400
Solomon	55			$300
Old Tom	70			$150
Tom Neman	48			$500
Tom Little	55			$300
Trim	68			$300
Tommy	76	cripple		$100
William	42	cooper		$1,500
York	46			$1,100
Females				
Aimee	26	and her children Marie, age 6 yrs & Eugene, age 4 yrs	2	$1,700

Appendix D

Slaves names	Age	Notes	# of Children	Value
Annette	55	a cook		$600
Becker	22	sickly—her children Zilia, age 3 and Robinson, age 2 months	2	$1,000
Betsy	55			$300
Claire	47			$700
Delphine	30	and her children Victor, age 12; William, age 10; Cesar, age 7; Bertelle, age 3; and Noel, age 4 months	5	$3,000
Elsin	18			$1,100
Elvire	45	sickly—and her children James, age 8 and Edward, age 4	2	$1,200
Francis	40	dropsical—and her children Margaret, age 15; Ellen, age 13; Harriet, age 11; Solomon (affected with hernia) age 9; Tom Jr., age 7; and Susan, age 5	6	$3,100
Frosine	70			$50
Fanny Simon	48			$400
Louis Little	16			$900
Denis	12			$700
Genevieve	40	and her children Cathiche, age 10; Augus, age 8 yrs; Marie Louise, age 6; and Rosalie, age 3	4	$1,600
Justine	24	and her child Auguste, age 2	1	$1,250
Henry	40	and her children Moses with one arm, age 16 and Nelson, age 12	2	$1,200
Iris	19	and her child	1	$1,200
Isabella	26			$1,100

Hermitage Slave Inventory of 1858

Slaves names	Age	Notes	# of Children	Value
Judith	60			$100
Dolly	12			$450
Louisa	27			$1,100
Louise	25	and her children Anderson, age 9; Ursin, age 6; and Casimire, age 3	3	$2,050
Letty	26			$1,100
Eliza	36	and her child	1	$1,000
Lucy	40	washerwomen—and her son	1	700
Cassus	15			900
Marsalla	32	and her children Anthony, age 11; Arthur, age 8; and Gabriel, age 18 months	3	$2,050
Mary	21	and her child	1	$1,150
Mary New	21			$1,000
Mary Little	45	affected with hernia		$400
Pauline	14			$750
Mary Norman	blank	crazy		0
Maria	45	and her children Juan, age 16; Francoise, age 14; and Martial, age 11	3	$2,500
Margaret	15			$1,000
Nanerine	26	affected with hernia—and her child	1	$1,000
Nancy	48			$450
Nelly	35	and her children Jacob, age 16 and Jane, age 14	2	$500
		(Jacobs Price)		$800
Phillis	22	and her children Martha, age 5 and Emilly, age 7 months	2	$1,500

Appendix D

Slaves names	Age	Notes	# of Children	Value
Patience	25	and her child	1	$1,200
Philomine	16			$1,100
Ruthy	52	sickly		$50
Rachel	48			$300
Sarah	22			$1,100
Big Sarah	24			$1,100
Silvy	55	and her child	1	$450
Anterine	14	(age 15)		$1,000
Sukey	20			
Sally	40	affected with hernia		$400
Old Sally	75			0
Sophy	58			$200
Vinah	58	sickly—and her son	1	$300
Jackson	17			$1,000
Robert	14			$750
Fanny Little	8			300
Totals				$107,700
Adult Slaves				113
Children				45
Slaves				158
Average Value				$682

* This inventory of Hermitage slaves in 1858 is a composite of two inventories (one in French and one in English) taken that year by the Bringiers. The only difference in the number of slaves on each involved the slave Fed who was not listed on the English copy.

Age Ratios of Hermitage Slave Force

Age Brackets	Number of Males	% of Male Slave Population	% of Entire Slave Work Force*	Number of Females	% of Female Slave Population	% of Entire Slave Work Force*	Total Number of Slaves	% of Entire Slave Work Force*
0 to 9 years*				1	2%	1%	1	1%
10 to 19 years	2	3%	2%	10	19%	9%	12	11%
20 to 29 years	14	24%	13%	17	31%	15%	31	28%
30 to 39 years	13	22%	12%	4	7%	4%	17	15%
40 to 49 years	10	17%	9%	12	22%	11%	22	20%
50 to 59 years	11	19%	10%	6	11%	5%	17	15%
60 to 69 years	3	5%	3%	1			4	4%
70 and older	2	3%	2%	2	4%	2%	4	4%
unknown				1	2%	1%	1	1%

Appendix D

*Children Ages	Males	Females	Total Number of Children
years less than 1	2	1	3
1	1		1
2	1		1
3		4	4
4	2		2
5		2	2
6	1	2	3
7	2		2
8	3		3
9	2		2
10	1	1	2
11	3		3
12	2		2
13	1		1
14	1	1	2
15	1		1
16	3		3
unknown			8
Total			45

Appendix E

Slave Names on Bringier Plantations

Trist Wood in his collection of materials on the Bringier family makes relatively few comments on the many slaves who spent their lives toiling for the family. However, in his miscellaneous notes, he states that in the documents he reviewed for the years 1858 to 1860 he found twenty-one separate rosters or lists of slaves for the Bringier plantations of the Hermitage, Houmas, and Brulé. As most of the list he used to collect his data are no longer available, his observations, summarized here, present an interesting peek into the identities of the Bringier's bondsmen.

According to Wood, the slave listings usually divided the bondsmen according to sex and groupings of adults and children. Some lists focused on the skills possessed by the slave—mason, engineer, cooper, etc. He found that the great majority of names were in English, with most bearing a single, simple Christian name such as John, Joseph, Peter, Philip, Jesse, Moses, Dick, and Abraham. Others had names associated with historic figures, Washington, Napoleon, Madison, Nelson, Monroe, etc. Some were more fanciful, Grandison, Quinzy, Sythes, etc. and others were from the classics, Caesar, Cyrus, and Pompey.

He found that approximately a quarter of the identities included descriptive terms for differentiation between slaves with the same common name, such as Allen Old or Old Allen, and Allen New; Bill and Bill Elder; Davis and Davis Brute; Tom and Tom Brown; Nathan Yellow and Nathan Black; Big Washington and Washington Little. Another common descriptive name was the term "buck", which slave owners often used to describe strong, virile male slaves, such as Billy Buck, Buck Billy, and Davis Buck. Other slave names given by the Bringiers related to the task assignments of the bondmen such as Simon Cooper and Sam Cook and some, such as Snooder and Mary Rubotton were names of unknown origin.

Across the South there was little consistency in the use of slave surnames. In some areas, such as the coastal regions of South Carolina and Georgia, the practice of slave surnames were more common than in other areas of the South. The Bringiers showed little consistency on the issue. Most slaves owned by the family had either simple names as noted above or had double names for descriptive purposes. However, there were a number of Bringier slaves who were listed with surnames. No explanation as to why some had surnames and others did not was offered either by Wood or found in the collection's documents. A sampling of these include Hampton Turner, Lewis McCargo, Washington McGuines, Hazel Williams, Henry David Ingram, Tom Sawyer, Henry Compton, Thomas Chisel, Stephen Wooden, Grand George, Tricky Trum, Mary Rubotton, Mary l'Hospital and Gustus Benjo.

The names of the family's female slaves generally followed the same patterns as the males. There were many simple names, including Mary, Sarah, Marthy, Lucy, Milly, Rachel, Janey, Jane, Phoebe, Delphine, Priscilla, Medora, Sylvie, Sophia, Arabella, Celia, Antoinette, and Melinda. As was the case with the males, female slaves also frequently were given descriptive names such as Malinda Yellow, Mary Ann Little, and Mary l'Hospital. The children on the plantations were given names similar to those of their parents, yet in no case did Wood find a child who was given a name that identified his/her parentage. See Bringier Papers, Miscellaneous Notes, 11-12.

APPENDIX F

Age and Sex Divisions of the largest Slaveholdings in Ascension Parish in 1860

Slaveholder	Size of Holding	Males 15-59		Females 15-59		Males Under 15 and Over 59		Females Under 15 and Over 59	
Bringier, L. A.	144	49	34%	42	29%	30	21%	23	16%
Bringier, M. S.	386	110	28%	134	35%	106	27%	36	9%
Burnside, John	753	274	36%	146	19%	207	27%	126	17%
Kenner, D. F.	473	203	43%	119	25%	74	16%	77	16%

Appendix G

*Bringier Plantation Ownership**

Plantation	Parish	Years	Owner
Ashland	Ascension	to 1880	D. F. Kenner
Ashland	Ascension	1881	D. F. Kenner and J. L. Brent
Ashland	Ascension	1890	Belle Helene Plts. Co. Ltd.
Bowden	Ascension	to 1858	H. B. Trist
Bowden	Ascension	1858–1868	not listed among active plantations
Bowden	Ascension	1869–1880	D. F. Kenner
Bowden	Ascension	1881–1889	D. F. Kenner and J. L. Brent
Bowden	Ascension	1890	Belle Helene Plts. Co. Ltd.
Brulé	Ascension	to 1853	Mrs. M. D. Bringier
Brulé	Ascension	1854–1872	no information listed, production figures included with those of Houmas Plantation
Brulé	Ascension	1873–1875	Benjamin Tureaud & Comp.
Brulé	Ascension	1876 onward	not listed among active plantations
Fashion	St. Charles	to 1851	R. Taylor
Fashion	St. Charles	1851–1862	R. Taylor
Fashion	St. Charles	1863–1870	information not available
Fashion	St. Charles	1870	Estate of E. W. Burbank
Hermitage	Ascension	to 1858	Mrs. M. D. Bringier
Hermitage	Ascension	1859–1861	L. A. Bringier
Hermitage	Ascension	1862–1868	information not available
Hermitage	Ascension	1869–1877	M. D. Bringier & Company
Hermitage	Ascension	1878–1880	L. A. Bringier & Company
Hermitage	Ascension	1881–1883	D. F. Kenner & L. A. Bringier

Hermitage	Ascension	1884–1887	D. F. Kenner
Hermitage	Ascension	1888	Estate of D. F. Kenner
Hermitage	Ascension	1889	Maginnis & Nolan
Houmas	Ascension	to 1862	Mrs. M. D. Bringier
Houmas	Ascension	1863–1869	information missing
Houmas	Ascension	1869–1880	Benjamin Tureaud & Comp.
Houmas	Ascension	1881–1885	D. F. Kenner and Benjamin Tureaud
Houmas	Ascension	1886–1887	D. F. Kenner
Houmas	Ascension	1888 onward	not listed among active plantations
Union	St. James	to 1862	Mrs. Tureaud
Union	St. James	1863 - 1869	information not available
Union	St. James	1869	Bartel and Jacobshagen

*See P. A. Champomier, *Statement of the Sugar Crop Made in Louisiana* (1844-1862) and Alceé Bochereau, *Statement of the Sugar and Rice Crops Made in Louisiana* (1879-1887).

APPENDIX H
Bringier Genealogy and Plantations

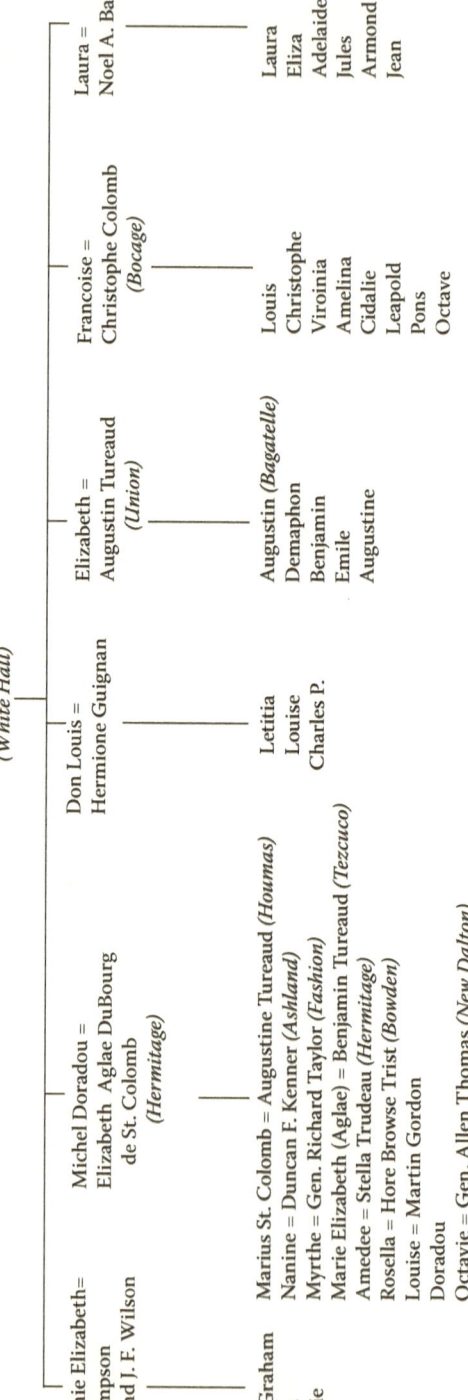

Bibliography

American State Papers, House of Representatives, 12th Congress, 1st Session, Public Lands: Vol. 2.

Banks, William Nathaniel. "The River Road Plantations of Louisiana." *Antiques* 3 (June 1977): 1170 - 1183.

Barck, Oscar T., Jr., and Hugh T. Lefler. *Colonial America*. 2nd ed. New York: The Macmillan Company, 1968.

Baudier, Roger. *The Catholic Church in Louisiana*. New Orleans: Roger Baudier, 1939. Reprint, Louisiana Library Association, 1972.

Bauer, Craig A. "The Last Effort: The Secret Mission of the Confederate Diplomat, Duncan F. Kenner." *Louisiana History* 22, no. 1 (Winter 1981): 67-95.

———. *A Leader Among Peers: The Life and Times of Duncan Farrar Kenner*. Lafayette, La.: Center for Louisiana Studies, 1993.

———, and Mark LaFlaur. "Duncan F. Kenner." *Dictionary of American National Biography*. New York: Oxford Press, 2002.

Bergeron, Arthur W., Jr. *Guide to Louisiana Confederate Military Units 1861-1865*. Baton Rouge: Louisiana State University Press, 1989.

Blassingame, John W., *The Slave Community: Plantation Life in the Antebellum South*. New York: Oxford University Press, 1972.

———, ed. *Slave Testimony: Two Centuries of Letters, Speeches, Interviews and Autobiographies*. Baton Rouge: Louisiana State University, 1877.

Boles, John B. *The South Through Time: A History of an American Region*. Englewood Cliffs, NJ: Prentice Hall, 1995.

Bouchereau, Alceé. *Statement of the Sugar and Rice Crops in Louisiana*. New Orleans: A. Bouchereau, 1878.

Bourgeois, Lillian C. *Cabanocey: The History, Customs and Folklore of St. James Parish*. New Orleans: Pelican Publishing Company, 1957.

Bringier-Colomb Family in Louisiana 1752-1943, genealogical chart compiled by Amedee Colomb, Sr., and revised by Clifford Colomb, July 1943, provided to the author by Duke Rivet.

Calhoun, Robert Dabney. "A History of Concordia Parish Louisiana." *Louisiana Historical Quarterly* 15 (January 1932): 44-67.

Champomier, Pierre A. *Statement of the Sugar Crop Made in Louisiana* (1844-1861). New Orleans: Cook, Young, and Company, 1845-1862.

Christovich, Mary Louis, Sally Kittredge Evans, and Roulhac Toledano. *New*

Orleans Architecture. Vol. 5, *The Esplanade Ridge*. Gretna, La.: Pelican Publishing Company, 1977.

Clinton, Catherine. *The Plantation Mistress: Women's World in the Old South*. New York: Pantheon Books, 1982.

Conrad, Glenn R., ed. *A Dictionary of Louisiana Biography*. New Orleans: Louisiana Historical Association, 1988.

———, and Ray F. Lucas. *White Gold: A Brief History of the Louisiana Sugar Industry 1795-1995*. Lafayette, La.: Center for Louisiana Studies, 1995.

Cooper, William J., and Thomas E. Terrill. *The American South: A History*, 2nd ed. New York: The McGraw-Hill Companies, Inc., 1996.

Daily Delta (New Orleans), 1861.

Daily Picayune (New Orleans), 1877-1878.

Daspit, Fred. *Louisiana Architecture: 1714-1830*. Lafayette, La.: Center for Louisiana Studies, 1996.

Davis, Edwin Adams. *Louisiana: A Narrative History*. 3rd ed. Baton Rouge: Claitor's Publishing Division, 1971.

Dufour, Charles L. *Ten Flags In the Wind: The Story of Louisiana*. New York: Harper and Row, 1967.

Durel, Lionel C. "Creole Civilization in Donaldsonville, 1850, According to 'Le Vigilant,'" *Louisiana Historical Quarterly* 31 (1948): 981-994.

Fogel, Robert William, and Stanley L. Engerman. *Time on the Cross: The Economics of American Negro Slavery*. Boston: Little, Brown and Co., 1974.

Follett, Richard. *Documenting Louisiana Sugar 1845-1917*. Sussex, England: University of Sussex, 2008, [database on-line].

———. *The Sugar Masters: Planters and Slaves in Louisiana's Cane World, 1820-1860*. Baton Rouge: Louisiana State University Press, 2005.

Foner, Eric. *Reconstruction: America's Unfinished Revolution, 1863-1877*. New York: Harper and Row Publishers, 1988.

Fuller, R. Reese. "Perique the Native Crop." *Louisiana Life* 23, no. 1 (Spring 2003): 44-49.

Gazette de la Louisiane, 1820.

Genovese, Eugene D. *Roll, Jordon, Roll: The World the Slaves Made*. New York: Pantheon Books, 1974.

Hair, William Ivy. *Bourbonism and Agrarian Protest: Louisiana Politics 1877-1900*. Baton Rouge: Louisiana State University Press, 1969.

Bibliography

———. *The Kingfish and His Realm: The Life and Times of Huey P. Long,* paperback edition. Baton Rouge: Louisiana State University Press, 1996.

Harrington, Fred Harvey. "A Peace Mission of 1863." *The American Historical Review* 46 (October 1940): 76-86.

Hermitage Foundation Papers. Hermitage Foundation, Darrow, Louisiana.
Trist Wood Papers

The Historic New Orleans Collection, Williams Research Center, New Orleans, Louisiana.
Robert Judice Collection
Trist Family Papers, Mss. 180

Hocker, Susan E. "Index to Early Louisiana Patents 1810-1890." Miami, Ohio: Miami University Libraries, 2002. http://staff.lib.muohio.edu/~shocker/LAPAT/invent.php?iname=Bringier (accessed March 27, 2007).

Johnson, Ludwell H. "The Red River Campaign." In *The Civil War Battlefield Guide,* edited by Frances H. Kennedy. Boston: Houghton Mifflin Company, 1990.

Johnson, Walter. *Soul by Soul: Life Inside the Antebellum Slave Market.* Cambridge, Mass.: Harvard University Press, 1999.

Kemp, John. *New Orleans: An Illustrated History.* Woodland Hills, Calif.: Windsor Publications, Inc., 1981.

King, Grace. *Creole Families of New Orleans.* New York: The MacMillan Co., 1921.

Kolchin, Peter. *American Slavery: 1619-1877.* New York: Hill and Wang, 1993.

Lachance, Paul. "Marriage and Property in Antebellum New Orleans." Paper presented at annual conference of the Association of Caribbean Historians, Havana, Cuba, April 15, 1999.

Laussat, Pierre Clement de. *Memoirs of My Life.* Edited by Robert D. Bush. Translated by Sister Agnes-Josephine Pastwa. Baton Rouge: Louisiana State University Press, 1958.

Louisiana State University, Louisiana and Lower Mississippi Valley Collections, Hill Memorial Library, Baton Rouge, Louisiana.
Ashland Plantation Record Book, Mss. 534
Louis A. Bringier Papers, Mss. 43, 139, 544
Duncan Farrar Kenner Papers, Mss. 198, 1402, 1477
Benjamin Tureaud Family Paper, Mss. 427, 560, 794, 811, 1100

Lewis, Peirce F. *New Orleans–The Making of an Urban Landscape.* Cambridge, Mass.: Ballinger Publishing Company, 1976.

Malone, Paul, and Lee Malone. *Louisiana Plantation Homes: A Return to*

Splendor. Gretna, La.: Pelican Publishing Company, 1996.

Marchand, Sidney A. *The Flight of a Century (1800-1900) In Ascension Parish Louisiana.* Donaldsonville, La.: N.p., 1936.

Melville, Annabelle M. *Louis William DuBourg: Bishop of Louisiana and the Floridas, Bishop of Montauban, and Archbishop of Bescancon, 1766-1818.* Vol. 1, *Schoolman, 1766-1818.* Chicago: Loyola University Press, 1986.

Menn, Joseph Karl. *The Large Slaveholders of Louisiana–1860.* New Orleans: Pelican Publishing Co., 1964.

Merrill, Horace Samuel. *Bourbon Leader: Grover Cleveland and the Democratic Party.* Boston: Little, Brown and Company, 1957.

Mitchell, William R., Jr. *Classic New Orleans.* New Orleans: Martin-St. Martin Publishing Company, 1993.

"Mother in the Civil War: The Wartime Reminiscences of Miss Wilhelmine Trist." Trist Wood Papers #800, Southern Historical Collection, The Wilson Library, University of North Carolina Library, Chapel Hill, North Carolina.

New Orleans Picayune, 1909.

Parrish, T. Michael. *Richard Taylor: Soldier Prince of Dixie.* Chapel Hill: University of North Carolina Press, 1992.

Peña, Christopher G. *Touched By Fire: Battles Fought in the Lafourche District.* Thibodaux, La.: C. B. P. Press, 1998.

Poesch, Jessie J. "Furniture of the River Road Plantations in Louisiana." *Antiques* 3 (June 1977): 1184-1185.

———, and Barbara SoRelle Bacot, eds. *Louisiana Buildings 1720-1940: The Historic American Buildings Survey.* Baton Rouge: Louisiana State University Press, 1997.

Prichard, Walter. "Routine on a Louisiana Sugar Plantation under the Slavery Regime." *Mississippi Valley Historical Review* 14 (September 1927): 168-178.

Remini, Robert V. *The Battle of New Orleans: Andrew Jackson and America's First Military Victory.* New York: Viking, 1999.

Rodrigue, John C. *Reconstruction in the Cane Fields: From Slavery to Free Labor in Louisiana's Sugar Parishes 1862-1880.* Baton Rouge: Louisiana State University, 2001.

Roland, Charles. *Louisiana Sugar Plantations During the American Civil War.* Leiden, Netherlands: E. J. Brill, 1957.

Russell, William H. *My Diary, North and South.* London: Bradley and Evans, 1863.

Bibliography

Scarborough, William Kauffman. *Masters of the Big House: Elite Slaveholders of the Mid-Nineteenth-Century South*. Baton Rouge: Louisiana State University Press, 2003.

———. *The Overseer: Plantation Management in the Old South*. Baton Rouge: Louisiana State University Press, 1966.

Seebold, Herman Boehm de Bachellé. *Old Louisiana Plantation Homes and Family Trees*. New Orleans: Herman Boehm de Bachellé Seebold, 1941.

Sexton, Richard. *Vestiges of Grandeur: The Plantation of Louisiana's River Road*. San Francisco: Chronicle Books, 1999.

Silliman, Benjamin. "Letter From L. Bringier, Esq. Of Louisiana to Elias Cornelius." *The American Journal of Science and Arts* 3 (1821): 19-20.

Sitterson, J. Carlye. *Sugar Country: The Cane Sugar Industry in the South, 1753-1950*. Lexington: University of Kentucky Press, 1953.

Smith, Frazer. *White Pillars: Early Life and Architecture of the Lower Mississippi Valley Country*. New York: Bramhall House, 1941.

Sternberg, Mary Ann. *Along the River Road: Past and Present on Louisiana's Historic Byway*. Baton Rouge: Louisiana State University Press, 2001.

Taylor, Joe Gray. *Louisiana Reconstructed: 1863-1877*. Baton Rouge: Louisiana State University Press, 1974.

———. *Negro Slavery in Louisiana*. Baton Rouge: Louisiana Historical Association, 1963.

Taylor, Richard. *Destruction and Reconstruction: Personal Experiences of the Late War*. D. Appleton and Company, 1879. Reprint, edited by Charles P. Roland. Waltham, Mass.: Blaisdell Publishing Company, 1968.

Thomason, Henry. "Ancient History Interpreted at Toltec Mound." Arkansas Department of Parks and Tourism, October 1, 2002. http://www.arkansasmediaroom.com (accessed March 8, 2007).

Tregle, Joseph G., Jr. *Louisiana in the Age of Jackson: A Clash of Cultures and Personalities*. Baton Rouge: Louisiana State University Press, 1999.

———. "On That Word 'Creole' Again: A Note." *Louisiana History* 23 (Spring 1982): 193-198.

———. "Thomas J. Durant, Utopian Socialism, and the Failure of Presidential Reconstruction in Louisiana." *Journal of Southern History* 45 (November 1979): 485-512.

Toledano, Roulhac, and Mary Louis Christovich. *New Orleans Architecture*. Vol. 6, *Faubourg Treme and the Bayou Road*. Gretna, La.: Pelican Publishing Company, 1980.

United States v. Watkins, 97 US 219 (1877).

Vogt, Lloyd. *New Orleans Houses: A House-Watcher's Guide*. Gretna, La.: Pelican Publishing Company, 1992.

Yakubik, Jill-Karen, and Rosalinda Mendez. *Beyond the Great House: Archaeology at Ashland-Belle Helene Plantation*. Baton Rouge: Louisiana Department of Culture, Recreation and Tourism–Division of Archaeology, nd.

The War of the Rebellion: A Compilation of the Official Records of the Union and Confederate Armies. Washington, D.C.: United States Goverment Printing Office, 1880-1901.

Williams, T. Harry. *Huey Long*. New York: Alfred A. Knopf, 1969.

Wilson, James D., Jr. "'To Live Together in Peace:' The Extraordinary Life and Times of Pierre Caliste Landry." Unpublished manuscript.

Wilson, Samuel Jr., and Bernard Lemann. *New Orleans Architecture*. Vol. 1, *The Lower Garden District*. Gretna, La.: Pelican Publishing Company, Inc., 1971.

Winters, John D. *The Civil War in Louisiana*. Baton Rouge: Louisiana State University, 1963.

Index

Aimee (slave), 193
Aleck (slave), 192
Alexandria, La., 110; *Democrat*, 106
Alexie, Ernest, 73
Alfred (slave), 87, 192
Allen, Nicholas, 73
Alluvial Land Purchase Company, 146
Amelung Plantation, 44
American Revolution; and education, 33
Amite River, 24
ancienne, 26
Andry Plantation, 14
Anduse, L' Abbé, 14
Annette (slave), 194
Anterine (slave), 196
Anthony (slave), 192
Antoine (slave), 90, 192
architecture; Classical-Revival style, 43; Creole plantation house, 42; early Louisiana, 4; Louisiana-Classic style, 43
Armant Plantation, 147
Arthur, Chester A., 132
Ascension Catholic Church, 49
Ascension Parish, ix-x, 14, 17, 19, 29, 37, 41-42, 50, 60, 64, 70-71, 98, 107, 115, 122, 133, 201
Ashland Plantation, x, 55, 60, 64, 66, 70, 80, 91, 108-9, 113, 115, 122, 130-31, 177-91; ownership of, 202; sugarhouse, 62, 161; Union raid on, 109

Audubon, John James, 8, 151
Augustin (schooner), 65
Augustin (slave), 3
Augustine (slave), 192
"Aunt" Carmelite (slave), 73
"Aunt" Ellen (slave), 73
"Aunt" Emma (slave), 73
"Aunt" Jane (slave), 73

Baconais, Estelle, 33
Baconais, Fanny, 33
Bagatelle Plantation, x-xi
Baltimore, Md., 21, 25-26, 28-29, 34, 47, 55
Banks, Nathaniel P., 96-97, 101-2, 115, 117
Baptiste (slave), 192
Barlow, Samuel, 136
Baron, Francoise Adelaide, 23
Baron, Francoise Laure Bringier, 23
Baron, Jules, 23
Baron, Noel Auguste, Jr., 23
barreaux, 4
Bastien (slave), 192
Bateman, William, 127
Baton Rouge *Gazette*, 91
Bauduc, Joseph Theodore, 52
Bayou, Lafourche, 49, 112; Tensas, 13
Baxter, James, 115-17, 121
Becker (slave), 194
Bel Cheney Springs, 108
Belton, Helen, 21
Ben (slave), 192

211

Benjamin (slave), 192
Benjamin, Judah P., 96, 99-100
Big Sarah (slave), 196
Black Code. See *Code Noir*
Black Hawk, 139
Black River, 13
Blackwater Plantation, 133
Bob (slave), 86, 192
Bocage Plantation, x, 19, 21, 39, 146, 185
Boré, Etienne de, 7
Bouchereau, Alcee, 125
Boudreaux, Ursin, 73
bousillage, 4
Bowden Plantation, x, 66, 122, 145, 178-82, 185-91; ownership of, 202
Brady, Edwin P., 146
Bragg, Braxton, 55
Brecks, George, 91
Brent, Joseph Lancaster, 57, 105-6, 145, 189-91, 202
Bringier, Augustine Tureaud, 37, 139, 173
Bringier, Charles Pendleton, 153
Bringier, Elizabeth Aglae DuBourg, x, xii, 9, 11-12, 19, 27, 29, 32-35, 38-40, 42, 47, 50,-52, 55, 72, 89, 109, 137, 138-39, 160, 173, 178-84; death of, 56
Bringier, Elizabeth Melanie. See Wilson, Elizabeth Melanie Bringier Simpson
Bringier, Felicie Augustine, 173
Bringier, Francoise (daughter). See Colomb, Francoise "Fanny" Bringier
Bringier, Francoise Laure. See Baron, Francoise Laure Bringier
Bringier, Hermione, 18-19
Bringier, Letitia, 18, 153
Bringier, (Don) Louis, 13-20, 31, 50, 77
Bringier, Louis Amedee, 32, 34, 36, 62, 66-68, 72, 77, 83, 85, 87, 89-90, 93, 107, 109, 113, 120-22, 124-30, 134, 145, 167, 173, 182-84, 188-90, 201-2; and Civil War, 103-6; patents, 67, 126, 175
Bringier, Louise, 18, 50, 153
Bringier, Louise Elizabeth. See Tureaud, Louise Elizabeth "Betsy" Bringier
Bringier, M. S., Jr., 137, 139, 173
Bringier, Marie Anne "Nanene" Roudanez, 10
Bringier, Marie Catherine, 1
Bringier, Marie Francoise Durand, 1-3, 6, 9, 25, 74
Bringier, Marius Pons, xi, 1-8, 13-14, 19-20, 23-26, 41-42, 89-90, 173; second marriage, 10
Bringier, Marius Ste. Colombe "M. S.," 11-12, 34-36, 39-40, 66, 68, 72, 77, 79, 93, 99, 112, 145, 173, 201; patents, 38, 67, 175-76; post war recovery, 124-25
Bringier, Martin Doradou "Dadou," 34, 125, 129, 137-38, 173; and Civil War, 103, 105
Bringier, Michel Dourado "Doradou," x, xii, 8-9, 11, 15-17, 18, 23-26, 28-32, 34-35, 37, 41-42, 47, 50-52, 63, 66, 71, 77,

Index

80, 91, 153, 159, 173, 177-78; death of, 36, 54
Bringier mausoleum, 36, 173
Bringier, Paul Louis. *See* Bringier, (Don) Louis
Bringier Pulverizing Cultivator, 67, 126
Bringier, Rosella. *See* Trist, Rosella Bringier
Bringier, Stella, 11, 62, 89, 109, 112-13, 121, 126, 129, 173
Bringier, Trist, 37
Bringier, Vincent, 2, 149
Bringier's Cavalry, 105
briquette entre poteaux, 4, 45
Brulé Plantation, x, 37-38, 145, 178-80, 185-88; ownership of, 202; perique tobacco, 71; sugarhouse, 62
Bunell (slave),192
Burbridge, Joseph, 74, 78
Burnside, John, 38, 68, 99, 201
Butler, Benjamin, 115

Cabildo, 5
Cage, Albert Gallatin, 84
Camp Moore, 104
Canby, Edward, 55
Cantrelle, Michel, 25
Cassus (slave), 195
Carmouche, Pierre, 73
Carondelet, Francisco Luis, Baron de, 13
Carroll, "Aunt" Harriet, 73
Chalmette, La., 9, 30
Chapitoulas District. *See* Tchoupitoulas District
Charles (slave), 192
Chauveau, Capt. J., 30

Chompomier, P. A., 124
Christmas (slave), 85, 192
City of St. Elmo, La., 146-47
Claiborne, William C. C., 23
Cleveland, Grover, 134
Clinton, Catherine, 29
Code Noir, 89, 132
Colomb, Francoise "Fanny" Bringier, 14, 19-20, 21, 39, 178-85
Colomb, Henry Octave, 130
Colomb, L. Arthur, 173
Colomb, Louis Christophe, 5, 11, 14, 20-21, 153, 173, 177-78
Colomb Plantation, x
Confederate Provisional Congress, 99
Confederate States of America, 95, 98
Congo, Joe, 86-87, 193
Cottman, Joseph, 74, 87
Cottman, Thomas, 74, 87
cotton, 7
Cozzens, C. W., 114-15
Creole, 13, 152
Creole (steamboat), 65
Crescent Regiment, 103
crevasse, 53, 70, 159
crops, 7
Currey (slave), 192

Darrow, La., ix, 148
Davis, Jefferson, 55, 96, 98-101, 137
Dedé (slave), 36
Destruction and Reconstruction, 136
Dolly (slave), 195
Donaldsonville, La., 10, 20, 34, 36, 41, 43, 49-50, 110, 114, 117,

120, 130, 133-34, 141; *Le Vigilant*, 49; Union occupation of, 108-9, 112
d'Orleans, Louis Philippe, 8
drainage system, 61
DuBourg, Abbé William, 8, 17, 26-29, 34, 47, 53, 55, 154
DuBourg, Louis, 28
DuBourg, Pierre Francois, 23, 25-26, 28

E

East Baton Rouge Parish, 133
education, 33, 34; of girls, 27
Eliza (slave), 195
Emile (slave), 90
Emiline (slave), 90
Esplanade Ridge, 17

F

Factors, 65-66, 76
Farragut, David, 95, 107-8, 112
Fashion Plantation, x, 55, 101, 135, 178-84; ownership of, 202
Fed (slave), 192
Federal Bureau of Abandoned Properties, 114
Field, Seaman, 52, 54, 66, 159
fireside etiquette, 27
1st Louisiana Regiment (Union), 113
Fish, Hamilton, 55
flooding, 70
Florida Sugar Manufacturing Company, 130
Fontenot, Sister Helen, xiv
Fort Butler, La., 117
Fort St. Charles, La., 50
Fort Sumter, S.C., 95, 98
Fosdick, George, 35

4th Louisiana Cavalry, 104, 106
Frank (slave), 88, 192
French Revolution, 19-20, 27
Frederick (slave), 192
Frosine (slave), 87

G

Gaines, Myra Clark, 55
Garbriel, Dr., 73
George (slave), 85, 192
Georgetown University, 28-29
Gordon, Louisa Francoise Bringier, 36, 40, 54-55, 124-25, 160
Gordon, Martin, 8, 54, 159
Gordon, Martin, Jr., 35, 39-40, 54-56, 79, 95-96, 117, 125, 130, 139; as factor, 66, 76
Gordon, Mary Wilhemine Trist, 173
Grand Lodge of Masons, 26, 28
Grand Opera House, 51
Grant, Ulysses S., 101, 110-11, 136
Graves, George Washington, 122-23, 130
Gravier, Jean, 32
Greeley, Horace, 133
Guignan, Marie Josephine Hermione. *See* Bringier, Hermione
Guillemard, Gilberto, 5

H

Hampton, George, 74
Hampton, Wade, 11, 37
Hancock, Winfield S., 133
Helvetia Plantation, 147
Henry (slave), 192
Hermitage Plantation, ix-xi, 9, 11, 28, 30, 32, 41, 55, 145, 157, 178-

91; addition, 46; construction of, 42-45, 157; ownership of, 202; restoration of, 147, 157; slave inventory of 1858, 85, 87-88, 192-98; slave housing at, 83-84; sugarhouse, 62-63; sugar production, 69; Union occupation of, 114-15, 117; Union raid, 113-114
Hispaniola, 25
"hoe gangs," 60, 77
Hollywood Plantation, 131, 133-34
Honoré (slave), 192
Hood, John Bell, 55
Houmas House Plantation, 68
Houmas Plantation, x, 17, 19, 37-38, 145, 178-91; ownership of, 156, 203; slave housing at, 84; sugarhouse, 63; sugar production, 69
Howe, Judge, 57

Iberville Parish, xi, 34
indigo, 7
Isaac (slave), 91

Jack (slave), 193
Jackson (slave), 196
Jackson, Andrew, 8, 30-31, 55
Jackson, Andrew, Jr., 9
Jackson, Rachel, 8-9
Jackson, Thomas "Stonewall," 101
Jamaica, 25-26
Janvier (slave), 192
Jefferson (slave),192
Jefferson College, 34
Jerry (slave), 192

Jesse (slave), 85, 192
Joe (slave), 192
John (slave), 193
Johnson, Andrew, 120, 137
Johnston, Joseph E., 102
Jones, Peter, 87, 193
Joseph (slave), 87, 193
Juan (slave), 88
Judice, Robert, xiv, 147
Judice, Susan, 147
Judith (slave), 195

Kellogg, William Pitt, 136
Kenner, Duncan Farrar, xi, 34, 39, 55, 57, 60, 62-64, 66-68, 70, 91, 98-99, 106, 108, 113, 122, 124-26, 129-32, 134-35, 138-41, 145, 160, 173, 177-91; 201-3; death of, 143; diplomatic mission, 100; service to Confederacy, 99
Kenner, George D., 173
Kenner, Nanine Bringier, 34, 39, 55, 57, 60, 80, 98, 108-10, 112, 124-25, 129-33, 138-42, 160, 173
Kolchin, Peter, 87, 89

Lachance, Paul, 29
Lacombe, Madame, 27
Lafourche country, 16, 110
Lake Maurepas, 17, 37, 44
Lalaurie, Delphine Macarty de Lopez Blanque, 159
Lalaurie, Louis, 159
Landry, Pierre Caliste, 73, 78-79, 164
Lardner, Thomas, 73

Laussat, Pierre Clement de, 5-6
"lay-by" period, 61
Lea, James, 73
Lee, Robert E., 96, 101-2
Leed's Light Horse, 103
Letty (slave), 195
Lincoln, Abraham, 96, 98, 100
Little, Ben (slave), 192
Little, Fanny (slave), 196
Little, Mary (slave), 195
Little, Tom, 193
Little Sam (slave), 193
livestock, 61, 64, 109, 113, 121, 144
Long Hot Summer, The, 147
Louis (slave), 193
Louis, Jean, 87, 193
Louisa (slave), 87, 195
Louise (slave), 195
Louisiana Scientific Agricultural Association, 131
Louisiana State University and Agricultural and Mechanical College, 133
Louisiana Sugar Planters' Association, 66, 131
Lubbock, Francis Richard, 23
Lucy (slave), 195

M*adriers*, 4
Maginnis, W. D., 145
Maison Blanche, La, 3
Malone, James, 73
Mansfield, La., Battle of, 102
Margaret (slave), 195
Maria (slave), 88, 195
marriage ages, 29
Marsalla (slave), 195
Martial (slave), 193

Martinique, 1, 6-7, 9, 74, 92
Mary (slave), 195
Mary New (slave), 195
Mason, James, 100
Maury, Dabney, 55
Mayne, George A., 113
McDonald, John, 126
Melpomene, x, 18, 33, 51-57, 59, 66, 72, 79, 89-90, 107, 125, 136, 138-41, 144, 159
Michel (slave), 193
Minor, William, Jr., 110
Mississippi River, ix, 2-3, 14-17, 19-20, 29, 37, 41, 44, 49, 59, 61, 65, 70-71, 73, 95, 102, 107-8, 110, 112, 114, 117, 148
Mobile, Ala., 111
Monroe, John T., 95
Moore, Thomas Overton, 55, 103, 168
Montgomery Convention, 98
Murfreesboro, Tenn., Battle of, 104, 140

N*ancy* (slave), 195
Nanerine (slave), 195
Napoleon III, 100
Natchitoches, LA, 110
Nathan (slave), 193
National Register of Historic Places, 148
Ned (slave), 193
Nelly (slave), 195
Neman, Tom, 193
New Basin Canal, 135
New Charles (slave), 192
New Dalton Plantation, 55, 110, 133
New Hermitage (homestead), 146

Index

New Hope Plantation, 133
New Jesse (slave), 192
New Madrid earthquake, 15
New Orleans, La., x, 2, 14, 20, 22, 23-26, 31, 50; capture of, 107; *Daily Delta*, 98; *Daily Picayune*, 144; Faubourg St. Mary, 32, 51; *Times Democrat*, 56
New Peter (slave), 193
New Sam (slave), 193
Norman, Mary, 85, 88, 195
Normandy, France, 23

Old Pleasant (slave), 91
Old Sally (slave), 196
Old Tom (slave), 86, 193
Oliver, William, 74
Opelousas, La., 110
Order of Bolivar, 134
Osage, 15
overseer, 77; planters' sons as, 77

Pakenham, Edward, 30
Palmerston, Henry John, 100
Part, Pierre, 41
patents, 175-76
Patience (slave), 196
Pauline (slave), 195
Peabody, George, 136
Pemberton, John C., 102
perique tobacco, 71, 114
Perry (slave), 193
Perryville, Ky., Battle of, 104
Peter (slave), 193
Phillis (slave), 195
Philomine (slave), 196
plantation store, 78
Poesch, Jessie, xiv

Polk, Leonidas, 103
Prevost, Francois Marie, 78
Price, Jacobs (slave), 195

Rachel (slave), 196
Randall (slave), 193
Raphael (slave), 193
Reconstruction plan, 120
Red River campaign, 97, 116-17
religion, 35
Reuben (slave), 193
rice, 7
Riverton Plantation, 115, 147
Rivet, Duke, xiv
Robert (slave; 22 years old), 193
Robert (slave; 14 years old), 196
"rolling season." *See* sugarcane harvesting
Roman Catholicism, 35
Roudanez, Marie Anne, 10
Roudanez, Nerestine, 10
Roudenez, Eugene, 73
runaway slave, 91
Runnymeade, 134
Russell, William, 49, 98
Ruthy (slave), 87, 196

St. Charles Hotel, 23-24
St. Charles Parish, x, 101, 135
St. Cloud Plantation, 130
St. James Parish, x, 3, 7, 10-11, 13, 23-25, 34, 37, 40, 71, 74, 145, 147; Cemetery, 10
St. John the Baptist Parish; 1811 insurrection in, 7
St. Landry Parish, 102
Sally (slave), 196
Sam (slave), 193

San Domingo, 10, 19, 21, 25, 27-28; slave revolt, 27
Sarah (slave), 196
Scarborough, William Kauffman, 74, 93
Schaubhut, Diana, xiv
Scott, John, 104
Scott's Cavalry, 103-4, 167
secession, 95, 97-98, 103
secondary cash crops, 61; tobacco, 71
Seton, Mother Elizabeth, 27
Seven Days campaign, 101
7th Louisiana Cavalry, 104
Shiloh, Tenn., Battle of, 103-4, 107
Shreveport, La., 49, 96-97
Silvy (slave), 90, 196
Simon (slave), 192
Simon (slave; carpenter), 193
Simpson, Blanche Kenner, 173
Simpson, William, 23
Sitterson, J. Carlye, 78, 123
slave drivers, 77
slave gangs, 77
slaves; at White Hall, 7; children, 88-89; clothing of, 81-82; diet of, 81; housing of, 82-84; incentives for, 79-80; medical care of, 87; names of, 199-200; punishment of, 91-92; purchase of, 84; value of, 85-87
Slidell, John, 100
Smith, Edmund Kirby, 96-97, 102
Solomon (slave), 193
Sophy (slave), 196
Southdown Plantation, 110
Sternberg, Mary Ann, x
sugarcane, 7, 59; grinding of, 64; harvesting of, 61-62; planting of, 60; processing of, 62; "strike", 65
sugar industry; improvements to, 66-67
Sukey (slave), 196
Sultana (steamboat), 31

Tangipahoa Parish, 104
Taylor, Myrthe Bringier, 40, 55-56, 101, 110, 135
Taylor, Richard, xi, 40, 55-56, 66, 96-98, 103-4, 117, 120, 124, 135, 137, 178-79, 178-84, 202; and Civil War, 101-2
Taylor, Zachary, 56, 101
Tchoupitoulas District, 2, 7, 13
Tensas Parish, 133
Territory of Orleans, 23, 25
Tezcuco Plantation, x, 37, 110, 171
3rd Rhode Island Cavalry, 117
31st Massachusetts Infantry, 117
Thomas, Allen, 55, 66, 102-3, 105, 107, 110, 120, 137, 173; post-Civil War, 133-34
Thomas, Allen, Jr., 110, 173
Thomas, Anne Octavie Bringier, 55, 102, 110-12, 134, 168, 173
Tilden, Samuel J., 136
tobacco, 61
Toltec Mounds, 15
Tommy (slave), 86, 193
Treaty of Ghent, 30
Trim (slave), 193
Trist, Augustine, 139
Trist, Bringier, 103, 127
Trist, Hore Browse, 39, 66, 92, 145, 160, 173, 177-82; 202

Index 219

Trist, Julien Bringier, 103, 140
Trist, N. B., 67, 175
Trist, Nicholas, 103, 130
Trist, Nicholas Philip, 55
Trist, Rosella Bringier, 35, 39, 173
Trist, Wilhelminia. *See* Wood, Wilhelmine Trist
Trudeau, George Zenon, 107, 113, 114
Trudeau, James, 107
Tureaud, Augustin Dominique, 21-23, 173
Tureaud, Augustine. *See* Bringier, Augustine Tureaud
Tureaud, Benjamin, 35, 110, 122, 125, 145, 174, 185-91, 202-3
Tureaud, Benjamin, Jr., 174
Tureaud, Emile, 104
Tureaud, George Mather, 104, 107
Tureaud, James, 104
Tureaud, Louise Elizabeth "Betsy" Bringier, 19, 22, 174
Tureaud, Marie Elizabeth "Algae," 80, 110, 125, 174, 179-84
28th Louisiana Volunteer Infantry, 104

Union Plantation, x, 23, 37, 145, 153, 179-84; ownership of, 203
United States Tariff Commission, 132
University of Virginia, 34

Vacations, 34-35

Vicksburg, Miss., campaign of, 102, 104, 110-11
Vieux Carré, 50, 51
Vinah (slave), 196
Varieties Theater, 51

Wade, W. G., 60
wage labor, 122
War of 1812, 17, 28, 30, 43
Warmoth, Henry Clay, 136
Warren, Rose, xiv
Washington, La., 110
Watermann, John R., 174
Watkins, John, 24
Wederstrandt, J. C., 41-42
White Hall Plantation, x, 3, 6-13, 22, 41, 67, 178-84; construction of, 5, 150; sugar production, 69
William (slave), 85, 193
Wilson, Elizabeth Melanie Bringier Simpson, 23-24, 174
Wilson, James Fisher, 24
Winters, John, 119
Wood, Robert Crooke, xii, 56, 111
Wood, Trist, xii-xiv, 11, 28, 32, 37, 40, 111, 156, 199
Wood, Wilhelmine Trist, xii, 56, 160
World's Industrial and Cotton Centennial Exposition, 132

York (slave), 193

Dr. Craig Bauer is a Professor of History at Our Lady of Holy Cross College in New Orleans, Louisiana. He is the author of *A Leader Among Peers: The Life and Times of Duncan Farrar Kenner* published by the Center for Louisiana Studies in 1993 and several journal and periodical articles on Louisiana history and the juvenile justice system. Professor Bauer is a charter member, past board member, and past-president of the Jefferson Historical Society of Louisiana. He is past holder of both the Freeport-McMoran Endowed Professorship in Business and Education and is the Nancy O'Neill Endowed Professorship III at Our Lady of Holy Cross College. He earned his doctorate in history at the University of Southern Mississippi and currently resides in Metairie, Louisiana, with his wife, Betsy, and daughter, Charlotte.

www.ingramcontent.com/pod-product-compliance
Lightning Source LLC
Chambersburg PA
CBHW030314080526
44584CB00012B/566